B R I N G I N G
L I F E
T O
E T H I C S

BRINGING
LIFE
TO
ETHICS

GLOBAL BIOETHICS
FOR A HUMANE SOCIETY

MICHAEL W. FOX

STATE UNIVERSITY OF NEW YORK PRESS

Cover photo was taken at the anti-WTO public demonstrations in Seattle, November 1999. Courtesy of the photographer, Eileen Weintraub.

Published by
State University of New York Press, Albany

For information, address State University of New York Press, 90 State Street, Suite 700, Albany, N.Y. 12207

Production by Diane Ganeles
Marketing by Fran Keneston

Library of Congress Cataloging-in-Publication Data

Fox, Michael W., 1937–
    Bringing life to ethics : global bioethics for a humane society / Michael W. Fox
        p. cm.
    Includes bibliographical references and index.
    ISBN 0–7914–4801–0 (alk. paper) — ISBN 0–7914–4802–9 (pbk. : alk. paper)
    1. Bioethics.   2. Environmental ethics.   3. Animal rights.   I. Title.

QH332 .F68 2001
174´.957—dc21   00–026522

10 9 8 7 6 5 4 3 2 1

Contents

Figures and Tables

Figures

Tables

Foreword

The life and work of Socrates eloquently evidence the degree to which even a relatively monolithic society, such as classical Athens, needs a gadfly to keep it honest in matters of ethics, political thought, and reflection on the good life. How much the more so then, does a society such as ours, riddled as we are with cleavages and moral dissension, require people to aim for a synthetic and synoptic moral vision, and to keep an eye firmly fixed on the Grand Prize, the well-being of the earth and all of its inhabitants.

This need for gadflies has grown all the more pressing as our ways of knowing and dealing with the natural world—science and technology—have become ever more fragmented and specialized, and our savants know more and more about less.

To make things even more vexatious, practitioners of science and technology, for most of the last century, have been captured by an ideology that distances science from engaging questions of value in general and ethics in particular, leaving these questions to be not only answered by society in general, but also articulated by the same amorphous entity, whose fragmented members have neither the background nor the inclination to understand the science in question, much less its moral implications.

The social response to mammalian cloning and its ethical implications provide a marvelous, but depressing example of the problem. Though the scientists involved in cloning Dolly were well aware of the unpreparedness of society in general to comprehend the facts of cloning, let alone its moral implications, they nonetheless cavalierly dropped the Dolly bombshell upon the fully unprepared public, a public that responded by declaring that cloning was against God's will, and whose primary mode of dealing with the issues was fear, a mode quickly manipulated and inflamed by theological opportunists.

Much of the ethics industry played to this emotion, and I doubt that more than a fraction of the public has any idea of what cloning is and what issues it raises. The same holds true all across biotechnology, a scientific and technological revolution more powerful than any we have ever before encountered.

Regrettably, as much as we need them, very few gadflies have emerged, for it is harder than ever for people to meet the qualifications for being a relevant gadfly. Unlike Socrates, one must master complex scientific changes in a variety of areas. In addition, one must see beyond the dazzling, promissory, glitzy rhetoric surrounding new technologies to underlying realities and not so obvious dangers, as well as also see beyond knee-jerk doomsaying.

Also, one must have more than a passing acquaintance with environmental science, biomedicine, ecology, international justice, agriculture, and economics. And finally, one must have a passionate and fearless commitment to the good of the planet and the life it supports.

This is no easy set of tasks in a world where animal activists battle environmentalists and each other; where people, whose central passion is rescuing Doberman Pinschers, wouldn't lift a finger to help an Airedale; where educated people pride themselves on their ignorance of science, and where too many scientists shamelessly declare that "ethics is just opinion." Yet, Michael W. Fox has shown himself equal to these tasks for the almost twenty-five years I have known him and his work. Having himself done important and original work in behavioral science and veterinary medicine, Fox is solidly grounded also in biotechnology and agriculture. Equally important, as I have often remarked, he is touched with the gift of prophesy, not in the apocalyptic Nostradamus sense, but in the Old Testament sense of one who can very early detect trends in society and extrapolate their full–blown implications. Things that Michael Fox predicted thirty years ago, which received ridicule, are today truisms and even profundities in the mouths of those who vilified him then.

The fate of gadflies is tenuous. Though Socrates himself affirmed that he should be fed at public expense in grateful response to his educational efforts, he in fact received his last meal from an ungrateful society. Fox, too, has been attacked from all sides, usually unfairly and by people who insist on killing the messenger, rather than engaging the issues.

But he has persevered in his moral campaign, and this book stands in eloquent testimony to his concerns. The reader may not agree with everything Fox says (and I suspect he would rather have it that way), but one thing is certain: One cannot fail to be stimulated by this book to think and reflect on what we might otherwise attempt to ignore. We are all the better for his efforts—humans, the animals, and the Earth we all share.

BERNARD E. ROLLIN

Acknowledgments

I would like to express my gratitude toward all my friends and teachers (both human and nonhuman) who have provided me with much of the material in this book. I am especially thankful for the focus and clarity that my wife Deanna Krantz has given me in my life and in this work. And I express my appreciation to Ellen Truong, who has put my words into manuscript form and into numerous helpful revisions thereof, and to Professors Bernard Rollin and Sharon Janusz for helpful critiques and suggestions.

Thanks also to my editor at SUNY Press, Dale Cotton, for his insight, skill, and sensitivity in helping me turn the original manuscript into a book that is more accessible to a wider readership than I could have ever hoped for.

Introduction

As individuals and as a society, we face a host of issues, the moral complexity of which necessitates a new way of addressing them and of finding solutions. We need a Life Ethic. When our ethical concerns are expanded beyond the traditionally narrow scope of considering purely human interests to include the entire natural world and its biotic community of plants and animals, we have what I term bioethics: an ethical worldview that is life-centered rather than exclusively human-centered. Bioethics also integrates animal rights and environmental ethics, which hitherto have been quite separate and often conflicting fields—and not adequately balanced with human rights and interests.

My view of bioethics has always been holistic, seeking to integrate various disciplines and spheres of concern. To bioethicist and oncologist Van Rensselaer Potter, I am especially grateful for identifying the integrative, interdisciplinary ("bridge") functions of bioethics with what he calls "deep" and "global" bioethics. While I have been educated and often inspired by the writings of many animal rights and environmental ethics-focused philosophers and advocates, I have not found any book that brings together these various spheres of concern and contemporary issues into a cohesive whole. Global bioethics facilitates this process, and the intent of this book is precisely that: to broaden the scope of ethics, of human enquiry and concern, and to use ethics as a bridge between various disciplines and pursuits in order to counter the various trends of specialization, reductionism, and academic isolationism.

Some readers may feel that I have not given sufficient critical attention to the writings of other ethicists in these various spheres, and to their particular school or orientation. For this I apologize. I am familiar with a considerable volume of literature dealing with

1

ethics and moral philosophy but much of it I have found to be too academic (rather than too scholarly—good scholarship is something else)—and of limited practical value. As a pragmatist, I see global bioethics as bringing ethics back to life—to our everyday lives. I believe that this is critical for a host of reasons that will be discussed in subsequent chapters.

My own brand of bioethics may not appear to come from any particular school of philosophy, so some ethical rationalists and empiricists may fault it for lack of scholarship, especially if I've not mentioned particular works they have read and liked, or have written themselves. But, in fact, the global bioethics that I seek to identify and promote is derivative of a metaphysical view (or visionary experience/intuition) of the sacred unity, integrity, and interdependence of all life. So yes, with its metaphysical inspiration, global bioethics bridges the spiritual and the material, the religious and the secular, and most importantly, I believe, takes ethics out of academia and the classroom and into our professional and personal lives.

One need not be a believer in God to accept the absolute ethic of reverential respect for life universal and life in particular. A biological and spiritual (biospiritual) basis for a humane and just society is arguably more acceptable and therefore attainable from a metaphysical realization of the sacred, than from a pragmatic assertion of altruism as being enlightened self-interest, or from the dogmatic proclamations of salvation-oriented faith traditions. But as theologian Teilhard de Chardin observed, "since all paths that rise, converge," it does not matter what path we take so long as we try to live in harmony with other sentient beings and endeavor to live by the Golden Rule and respect all creatures and Creation.

The metaphysics behind this global bioethics sees biological reality as spiritual revelation, and biological evolution as both adaptation, spiritual development, and fulfillment or self-realization. The fulfillment of being is linked with the will to flourish for all sentient life. That fulfillment is the natural right or entitlement of all beings under our dominion. In practical terms it means allowing and enabling pigs and dogs to experience the fullness of their being, their "pigness" and "dogness," which is problematic when we raise pigs in factory farms and keep dogs in laboratory cages.

The conceptual approach offered by bioethics provides a broad-based framework for related matters such as public policy, corporate responsibility, social justice, and environmental and animal protection, in a diversity of fields including agriculture, veterinary medicine, conservation, biomedical research, and biotechnology. When applied to these fields, the principles of bioethics will do much to minimize costs and risks and maximize the short- and long-term benefits to society. With an ethical compass, all human activities and aspirations can be better directed to maximize the greater good. Harm will be minimized by such sensitivity.

Scientists admit that they lack the tools to predict future outcomes even with the best computer modeling. While the compass of ethics does not allow us to see into the future either, it does enable us to better ensure that the outcome or consequences of human activities and aspirations will result in more good than harm. In sum, since we lack the power to predict the future with absolute certainty, an ethical compass is an important instrument to help us avoid causing future harm by our current actions and aspirations.

By incorporating bioethics into their business practices, the industrialist, developer, and investor will indeed enjoy profits both material and nonmaterial. And would not all good people prefer to pass on the knowledge that they have not mortgaged the future, in the name of progress and necessity, by limiting the options of generations to come—by wasting nonrenewable resources, annihilating wildlife habitat, polluting the environment, and defiling the beauty and sanctity of the natural world?

Van Rensselaer Potter first used the term bioethics in his 1971 book *Bioethics, Bridge to the Future*[1] to link biological science with ethics. In his 1988 book *Global Bioethics: Building on the Leopold Legacy,*[2] Potter shows how bioethics can serve as a bridge between disciplines, like ecology and medicine, as well as a bridge to the future. In this book, I expand this concept of bridging between disciplines and human activities, relationships, and aspirations. I include the welfare and protection of animals, wild and domesticated; developments in genetic engineering biotechnology and impacts of "free" trade and the world market; sustainable agriculture practices; and the conservation of biodiversity and cultural diversity. Building these bioethical bridges is more than an intellectual exercise—it is essential for our survival and for the future integrity of earth's creation and ecosystem. These bridges help link our personal and professional lives with those values, principles, and virtues that are synonymous with a humane society, with corporate accountability and responsibility, with an effective justice system, and with an educational system that helps realize students' creative potential and ethical sensibilities, inculcating a sense of personal responsibility and moral integrity. And these bridges must be cross-cultural.

By way of illustration, this bridging process connects rudimentary veterinary ethics (that have to do primarily with issues like professional conduct, approved drug use in farm animals, animal euthanasia, and other dilemmas in daily practices) with animal and environmental protection issues, public health, and medical ethics. It also attends to the more global issues of habitat and biodiversity conservation, minimization of the adverse impacts of domestic animals, and efforts to feed a hungry world. In his videotape script for the 1998 Fourth World Congress of the International Association of Bioethics in Tokyo, Potter asserts that the practical value of bioethics comes from "building *bridges* to *each* of the specialties and bridges *between* the specialties in order to further develop a Global Bioethic that sees human welfare in the context of respect for Nature." Potter perceptively advises that, "in present times, medical ethicists should go beyond monitoring technological fixes for the overprivileged."

We do not have scientific tools to predict the future accurately. But we can glimpse some of what the future will offer through the prism of our behavior and values—our ethics or lack thereof. These are shaping the world now for the future. Combining science and ethics as bioethics gives us foresight. To paraphrase George H. Kieffer, Ethics deals with what *ought to be* and provides a vision of the future in a way that contrasts with the present. Ethical decisions therefore help guide future actions in terms of future consequences.[3]

Our powers of dominion and instrumental knowledge must be tempered by the higher powers of humility and compassion. These powers are codified and expressed in the principles of bioethics and *praxis* of the Life Ethic, as developed in this book. *Ahimsa*, the doctrine of avoiding harm to other sentient beings, is central to living in accord with the Life Ethic. There is no need for violence (*himsa*) in human affairs, and whenever there is conflict, there are ethical criteria to accomplish resolution. This exposition of the Life Ethic is a preliminary starting point for further refinement and illumination.

No matter what our station, situation, or circumstances, the deeper significance and purpose of human existence is revealed when individuals and communities alike are centered on the Life Ethic. We live in a pluralistic society, with a diversity of views, values, and desires, a society that is rapidly becoming a global economic community where cooperation must take precedence over competition, and where both cultural and natural biodiversity must be valued and protected with the same vigor as individual human rights, the spirit of free enterprise, and self-determination. As this global community develops—a community that includes animals and natural ecosystems—the application of bioethics will help ensure that this development will lead to a humane, sustainable society for the benefit of generations to come.

I have given particular attention in this book to how our food is produced, and to the new technology of genetic engineering. This is because agriculture and genetic engineering biotechnology have profound social and environmental consequences for all generations to come and, therefore, warrant close scrutiny from a bioethical perspective. This book is for all who care and who want to make the world a better place. The power of ethics gives us increased control over our lives and enables us to make the right choices as consumers, voting citizens, and public or corporate policy decision makers. It is also a book for students who, in the course of their education and subsequent to their graduation, face many ethical questions, the outcome of which will determine their future success, security, and satisfaction.

A purely materialistic and self-centered mode of collective existence means a dysfunctional society and planetary ecosystem. It leads to a state of ecological anarchy that in turn means a non-viable economy and society in the long run. This mode was the aspired to "healthy" norm for our consumptive, competitive industrial society until the end of the second millennium A.D. Then climatologists began to reach unanimity over the "anthropogenic," human contribution to global warming, acid rain, and other atmospheric anomalies. By the 1980s, acid rain, a growing ozone hole over Antarctica and Australia, and mass famine in Africa were some of the many indicators that this mode of existence and the very principles of industrialism were as harmful as they were unethical.

Any discussion of bioethics, therefore, should begin from this perspective; otherwise, it is simply empty rhetoric. Likewise, any religion that does not include concern and reverence for nature and all sentient life is simply self-serving and ultimately sacrilegious. Thomas Gladwin observes:

> Since the Enlightenment, we have progressively differentiated humanity from the rest of nature, and separated objective truth from subjective morality. The greatest challenge of post-modernity may reside in their reintegration. A similar challenge may exist for theories of management.

Organizational science has evolved within a constricted or fractured epistemology, such that it embraces only a portion of reality. The organic, biotic, and intersubjective moral bases of organizational existence, we submit, have been neglected or repressed in the greater portion of modern management theory. This has resulted in theory which is at best limited, and at worst pathological. By disassociating human organization from the biosphere and the full human-community, it is possible that our theories have tacitly encouraged organizations to behave in ways that ultimately destroy their natural and social life support systems.[4]

The state of the environment and of the animal kingdom reflects how industrial society perceives nature and nonhuman life forms. It is evident that until there is a universal reverence for all life, human life itself will continue to be demeaned and harmed by a state of mind that is incompatible with the coexistence of anything that might resemble a natural world. How we value and relate to the rest of the earth's community (of subjects rather than objects) will determine the fate of that community, as well as our own destiny. The humane and effective prevention or control of diseases, pests, and our own population growth is part of the burden of responsibility that the gift and power of dominion, as planetary stewardship, imposes upon us all. This responsibility does not mean that the bioethical principle of reverence for all life need be abandoned when dealing with pests and pestilence, predators and parasites, for it is an ethic that can never be arbitrary or capricious. This ethic does not prohibit the taking of life or the control of species' numbers, but rather casts such necessary actions within a framework that will further the greater good with a minimum of harm.

It is no coincidence that the biosphere is now becoming as dysfunctional as the humanosphere, wherein nation-states are at war or experiencing socio-political and economic turmoil, and ecological devastation as industrial growth continues. It is an artifact of dualistic perception to separate the two. Humanity and nature are interdependent and interconnected; when one is harmed, so is the other. It takes empathy and common sense, not some leap of the imagination, to know that when we destroy the forests and pollute the oceans, we impoverish and harm ourselves; likewise, when we demean other animals or races, we do no less to ourselves.

When an ecosystem is badly damaged, as by declining biodiversity, there is no steady state, dynamic equilibrium and systemic homeostasis. The natural development for most living beings that are not pathogens likewise becomes abnormal in such an environment. Organisms become diseased when ecosystems are damaged; their population balance is disrupted. The more this organic whole is harmed, the more its myriad parts are harmed, relationships change, symbiotes become competitors or parasites, and species become extinct, as others—including humans—multiply, sicken and suffer.

We are slow to learn that society becomes dysfunctional, along with its economy, industry, and agriculture, when these organic interdependencies and the ethics that arise therefrom are disregarded. Without a fundamental change in our attitude toward the life community of this planet, all technological "fixes," scientific solutions, political panaceas, and economic recovery programs will fail. And the more we put our faith in such solutions, like the religious fundamentalists who espouse prayer and punishment, and the scions of industry who see progress purely as materialistic and without end, the more humanity will pull the rest of the world down with it. When the trees are all gone, the sky will fall. Our salvation and redemption begin with the Lakota Indian affirmation in thought and deed: *Mitakuye oyas'in*—we are all related.

A collectively arrogant chauvinism cannot conceive of a world where people hold all life in respect and reverence and give equal and fair consideration to all living beings. Such an egalitarian attitude respects not only the *intrinsic value* of living beings, but also their *extrinsic* or *instrumental value*, as distinct from their utilitarian value to humans. Thus the earthworm and bacteria in the soil are seen as playing such vital roles in ecology that their extrinsic value is greater than the sum of most of our own endeavors. This egalitarian, Earth- or Creation-centered attitude is the basic template of bioethics, which considers the life of the individual and the life of the whole in a non-dualistic way, so as to seek ways to sustain the well-being of humanity in harmony with the rest of creation. It should not be confused with the more simplistic (dualistic) and polemicizing rhetoric and philosophies of animal rights/liberation and Earth First! deep ecology. These areas of legitimate concern will gain greater public recognition and support when, through bioethics, they become part of a more universal agenda of achieving a humane and sustainable global economy and community. By the same token, the development of such a community and economy will never be achieved if animal and environmental concerns are excluded from the agenda and definition of community and sustainability. To be universal, one must be particular, mindful of both the whole and all its living parts.

Our current chauvinistic state of mind that can never say *enough* is the legacy of the conquistador mentality. Like the industrialists and missionaries who came after them, conquistadors tore apart the life communities of plants, animals, and indigenous peoples whom they discovered in the New World and incorporated into their colonial empires. Within this frame of mind the right solutions, that are so urgently needed today to save civilization and nature, will never be found. It is only with the right heart that the powers we possess can be best used to heal all our relations and relationships. An auspicious beginning would be for us to start protecting natural biodiversity with as passionate a commitment as we should have toward the preservation of cultural traditions, identity, and diversity, and toward upholding the right of every living being to be respected and given fair consideration. Nature must once again become our primary source of revelation and experience of the sacred, since the preservation of nature is our ultimate security and salvation. Where nothing is sacred, nothing is secure.

Cross-cultural and subcultural differences in values, beliefs, and cognitive processing have long been a source of conflict. The way in which people structure reality, their worldview, is to a large measure determined by cultural influences. And it is our worldviews that affect how we relate to each other, to foreigners, to animals, and to the environment.

There may be a lack of evident compassion. This may be cultural, but it is more likely to be situational. I have been to countries where the eyes of animals, in terms of emotional expression, seemed more human than the eyes of many of their keepers, who showed no emotion. Yet under the protective armor of cold indifference, under the despair and fatalism, there is a heart that still may be reached. But until the condition of people—their material and spiritual poverty—is addressed, they may remain closed to the appeals of the humanitarian and conservationist. Without hope, people will not be motivated to protect what is left of their environment. Without hope, they will not be moved to show compassion toward animals, or even each other. Poverty and hunger ultimately lead to a collapse of cultural values, thus creating an impasse to the acceptance of bioethical principles and praxis far greater than that caused by cross-cultural differences in values and worldview, particularly in the areas of conservation and humane treatment of animals. Animal cruelty and environmental desecration are normative for some societies and various institutions. But where there is economic security, negative cultural attitudes toward nature and animals are more amenable to transformation via education and legislation. As cultures and communities worldwide become more economically and socially interconnected, the role of bioethics in facilitating mutual recognition of interests and concerns is considerable.

It is an ironic but telling coincidence that under industrialism in the developed world and under the impetus of poverty in the Third World, all life is reduced to mere utility. There is no fundamental difference between poachers killing elephants and rhinos for their tusks and horns and a biomedical scientist killing dogs and monkeys. Both forms of animal exploitation are for human gain, and animal life is reduced to mere utility, a means to human ends. Likewise, there is no difference between a peasant community destroying the last trees for fuel wood and a timber company turning a forest into board-feet of lumber. Though the motivation may be different, the consequences and underlying attitude of utility and personal gain are the same, regardless of scale or species.

To be wholly preoccupied with one's own personal salvation, or material aggrandizement, is as unbalanced and self-limiting as to be focused *exclusively* on such issues as abortion, drug-related crime and violence, child abuse, and world hunger. We can neither insulate ourselves from these difficult times nor find effective solutions when we address only the symptoms and not the underlying causes of human violence and suffering. Either reaction—denial or inappropriate action—merely perpetuates these problems. It is a tragic fact that the best intentions of good people so often cause more harm than good—more harm collectively than those few of evil intent—because, lacking a holistic view of the causes of human violence and suffering, they address the symptoms instead of the causes. Sometimes, as in allopathic medicine, the focus is limited by a narrow, mechanistic concept of human well-being, of health and disease, diagnosis, and treatment. In other instances, the limitation is due to a constricted, moralistic, and judgmental bias, with often violent consequences in the name of justice or religious freedom.

The inherent limitations of situational ethics and moral relativism that can lead to the acceptance of evil means to achieve good ends are averted by applied bioethics. Likewise, the limitations of liberalism, which can undermine the greater good of society, can be circumvented. The antidote to both liberalism and totalitarianism is to give equal and fair consideration to the rights and interests of the individual and of society as a whole in order to ensure balance and harmony.

There are certain truths and values that all peoples embrace and upon which every just and sustainable society is built. These truths and values, as embodied in applied bioethics, can help us accommodate our cultural, religious, racial, and other diversities of opinion and belief. The basic principles and scope of bioethics will be discussed in the next two chapters. The first chapter is more of a personal reflection on the origin of our ethical sensibility that provides a basis for examining bioethics from a more "global" perspective.

The Origin of Ethics: Personal Reflections

ethics: 1. the discipline dealing with what is good and bad or right and wrong or with moral duty and obligation.
2. a group of moral principles or set of values.
3. character or the ideals of character manifested by a race of people

—*Webster's Third International Dictionary*

I was recently asked from what authority do my ethics come. This is a thought-provoking question. Some might say ethics comes from one's culture with its laws and moral codes, or from the higher authority of religious teachings. I replied that "my ethical sensibility comes from my feelings, awareness, and concern."

Ethics has been defined as "that brand of philosophy dealing with values relating to human conduct, with respect to the rightness or wrongness of certain actions and to the goodness and badness of the motives and ends of such actions."[1] Ethics from this definition calls for a close study of human behavior, of our values and goals, and also the consequences of how we choose to behave and live.[2]

Ethical behavior is not purely instinctual, though the capacity to behave altruistically is an inborn species characteristic shared with some other animal species. In order to be ethical, we must learn to become ethicists in childhood. This entails knowing how to evaluate impartially our own desires and behavior and those of others (including institutions and corporations) from the perspectives of: axiological ethics (examining values and beliefs), deontological ethics (identifying obligations and moral duties), and consequentialism

11

(considering the consequences of values and actions). Ethics must be incorporated into school curricula and embraced by families and communities if children are to become ethical and, in the process, actualize their humanity. To be inhuman is to be unethical. To be unethical as an individual, family, community, corporation, or institution, is to bring evil into the world.[3] By evil I mean that which is defined as morally wrong, immoral, harmful, injurious. Those who claim that predators must be evil because they harm and kill other animals for food, commit what I call the "zoomorphic fallacy." Nonhuman predators are amoral in this respect, true to their natures, and unlike humans, have no choice. Yet, natural predation is used to justify killing animals for "sport" on the one hand, and to argue that nature is flawed or "fallen" on the other—both views an extension of the zoomorphic fallacy.

Certainly our various cultural backgrounds and religious beliefs influence our ethical sensibility to a considerable degree. And recent studies of human twins raised separately in different environments indicate that there is also a significant genetic influence on the development of moral awareness and ethical sensibility. So it is logical to conclude that both Nature and nurture play a significant role in moral development and ethical sensibility. There are also clear sex-correlated differences, according to psychologist Carol Gilligan, in how boys and girls make moral decisions and deal with ethical issues. In her book entitled *In A Different Voice*, she contends that women tend to operate on an ethic of care, centering on the necessary interdependence among human beings. Their moral priority is "each person's responsibility to care for others."[4] In contrast, men tend to rely on "a justice approach" that focuses on the human desire for self-fulfillment. The moral imperative for men is "each person's right to be protected from the interference of others." These are what Gilligan describes as "different voices," but she is careful to qualify this by noting that they are gender-related rather than gender-specific. Most people speak in both voices, Gilligan believes, but there is a tendency to have a major and a minor mode. Men tend to think in terms of rights, while women tend to think in terms of responsibility. Clearly, it is necessary to hear both voices, and ideally there must be a balance between major and minor modes in each of us. More esoterically, the male, or yang, mode is one of considering freedom, rights, and the interests of self, while the yin, or female, mode thinks more in terms of kinship, selflessness rather than selfishness, and mutual obligations and responsibilities. The yin mode is nurturing, embracing, and conserving, as the Jungian archetype of the *anima*, while the male *animus* tends towards exploitation, separation, and change. These two polarities are not mutually exclusive, and yet must be reconciled. By analogy, these two polarities represent concern for the rights of the individual, animal or human, on the yang side, versus the more ecological, holistic dimensions of concern for the whole, i.e., ecological interrelationships on the yin. It is within this dialectical tension that the rights and freedom of the individual must be cast—that is, within a broader ecological and egalitarian framework of responsibility, mutual obligation, and respect. Lao Tzu, the sixth century B.C. philosopher, must have the last word here: "He who knows the masculine and yet keeps to the feminine will become a channel drawing all the world to it; being a channel of the world we will not be severed from eternal virtue."[5]

While some educators and contemporary political and religious figures contend that morality's foundation is religion, and that the essence of religion is morality (thus morality, politics, and religion are inseparable), the real foundation of morality and ethics is in our love and respect for all life. It is not so much the inability of fish and other creatures to communicate their sufferings to us that is the problem; it is our inability to understand and empathize with the suffering of others, be they fish, fowl, or even our fellow beings. A society that does not recognize this is in a state of disorder: hence the importance of humane education that does not impose a rigid, conformist morality upon children but instead leads them along the path of empathy and ethical sensibility from which conscience and moral sensibility arise spontaneously.

Children who crush insects and other small creatures under their feet, or pull them apart, may do so not simply to feel powerful and omnipotent or because they fear the creatures might harm them. Some may do it because they have been abused themselves, or identify with the vulnerability and smallness of such creatures. By eliminating such a lesser life form, children may be endeavoring to erase that which they fear in themselves, and are unable to accept: their own powerlessness, insecurity, helplessness, and vulnerability. Children can understand and overcome these unconscious reactions by being encouraged to identify, in their own smallness and vulnerability, with lesser creatures. In fact, many children do so spontaneously. Once they learn to empathize with such animals and recognize that they are alike in many ways, a great stride is made toward both self-acceptance and a reverence for all life. But cultural, age, and sex differences aside, it would seem that we have an instinct as conscious and empathetic beings to act with conscience, and that it is the unified sensibility of feeling and reason that is the basis of ethics.

Wanton destruction and wastefulness can also be interpreted as forms of violence and the antithesis of the Jain and Buddhist doctrine of ahimsa, as well as of the Christian virtues of frugality and love. The supreme ethic of love subsumes the active principle of doing no harm—even to the least of Creation.

Ethics—Situational and Absolute

One situation or context is always slightly different from another. Hence, what is right in one situation might be inappropriate in another context. In recent recognition of this phenomenon of ethical relativity, philosophers now speak of *situational ethics*. The ethic of nonviolence, for example, is not absolute. It is not possible to avoid acting violently in *all* situations, especially in some emergencies that threaten life or basic rights.

Situational ethics is used by instrumentalists, when they argue that the suffering of laboratory animals is justifiable because from it many sick people benefit (and jobs and profits come before environmental concerns). The situational nature of this argument is exemplified by the fact that dog owners would be prosecuted for doing what scientists do in the laboratory—burning a dog with a blow torch or poisoning him with pesticide. Logically, the dog is still a dog in both contexts, yet in one situation his suffering is condoned by society.

The danger, then, of situational ethics is that it allows violence to become a culturally acceptable norm in certain purportedly necessary and unavoidable situations. Clearly, it is incumbent upon us to question the source of our values and ethics and the ends they ultimately serve. Such questioning will surely help us to identify areas of ethical inconsistency and moral ambiguity that can have harmful consequences. This process enables us to reduce such harm in the future.

According to Australian aboriginal elder Bill Neidjie, *The Law*, as he calls it, to respect the environment and to waste nothing, is sacred. Those who break this law will suffer the consequences.[6] Native and aboriginal peoples regard wanton destruction and wastefulness as cardinal sins, whereas conservation and frugality are regarded as virtues derived from a reverence for Mother

Nature and from enlightened self-interest. It is from this ancient wisdom that Aldo Leopold, a federal wildlife biologist, derived his *land ethic* that holds that a thing is right when it tends to preserve the integrity, stability and beauty of the biotic community. Without a deep feeling for the land, which comes from being consciously part of it, concern for the environment becomes marginalized from public consensus and is ultimately excluded from the scope of moral concern.

The instinct for self-preservation becomes self-destructive when it is not linked with the basic human ethic of holding dear the survival of planet earth, as well as the survival of any living creature. But when two are related, the self-preservation instinct becomes creative rather than destructive. So linked, this instinct is the driving force of the conservation and "deep" ecology movements.

The second instinct, to avoid pain in oneself, when extended via empathy as compassionate concern for the suffering of others, becomes the humane ethic. This ethic is the force behind both the human and animal protection/rights movements. When not supported by empathy, this concern is often based on sentimentality and patronage. In the absence of the humane ethic, cruelty toward animals and unnecessary exploitation become acceptable cultural norms. And with the erosion of ethical sensibility toward all living things, comes acceptance of others' suffering in the search for ways to alleviate one's own pain. In sum, as concern becomes more and more self-centered, an ethical boundary is set up between humans and other beings where once there was the bridge of empathy and a source of kinship.

Ethical sensibility is inherent in all human beings, but it cannot arise without love and respect early in life, during which time the personal value of adhering to moral codes is learned. In adolescence, such learning is often questioned and tested, even challenged. Ultimately, obedience to moral codes is transcended in the process of maturation and personal liberation from social consensus and the ideological orthodoxies of the times. For a person's ethical sensibility to arise in a culture that is deficient in love and respect for all, such transcendence necessitates a courageous transformational *leap*. This leap results in the birth of radical consciousness, of ethical sensibility, and empathetic sensitivity. This leap of faith, philosopher Kierkegaard said, is one from the esthetic to the ethical. In *Radical Man* Charles Hampden Turner observes, "Before he leaps, man beholds the aesthetic. When he has leapt, the aesthetic can be consummated to become the ethical. In this way he can overcome the dichotomy between the subjective and the objective to resolve all polarities that originate in Cartesian dualism.[7]

Aesthetics and Ethics

Aesthetics comes from the Greek *aisthesis*, meaning perception. Aesthetic sensitivity, or how we feel about what we perceive, is the substrate for ethical sensitivity and for those values and actions which are consonant with what we recognize as beauty, harmony, vitality, freedom, and fulfillment. Aesthetics is the connecting link between ethics and how we perceive and value things, and influences our relationships with other living beings. Aesthetic criteria of value, as in concern for and delight in the fulfillment of other beings and nature, is the essence of illimitable love. And it is from this love that the supreme ethic of reverence for the sanctity of being and for the integrity of Creation arises. Through the feeling-awareness of empathy and aesthetic sensitivity, perception is clear and unbiased. Without these faculties, perception becomes clouded by one's own subjectivity and can lead to the anaesthetic state of unfeeling ethical blindness, with total insensitivity to others' suffering.

When the aesthetic sense is lost, as when one becomes habituated and then indifferent to urban squalor, factory farms, and the industrialized desecration of nature, ethical blindness is almost inevitable. While habituation and desensitization may be adaptive coping mechanisms, the loss of ethical sensibility and aesthetic values is ultimately maladaptive. Likewise, through lack of socializing contact early in life with animals and with nature, aesthetic appreciation for what is natural is seriously limited. The ability to discriminate the real and the unreal, the natural and the unnatural, is thus impaired, and again the end result is ethical blindness.

Beauty, Ethics, and Aesthetics

Our perception and appreciation of natural beauty may also be biological attributes of the human psyche meant to enhance survival. We are moved to respect and to protect that which we value for its inherent beauty. And the transpersonal experience of beauty can fill the human psyche with the numinous, spiritual dimension of the natural Creation. Aesthetic awareness may be the instinct that leads us to Divinity.

Poets, who are mindful of beauty, have repeatedly reminded us of nature's value. Robinson Jeffers wrote:

> The beauty of things—
> Is in the beholder's brain—
> the human mind's translation
> of their transhuman
> Intrinsic value.

He saw that:

> The greatest beauty is organic wholeness,
> the whole of life and things,
> the divine beauty of the universe.
> Love that, not man apart from that. . . .[8]

Environmental theologian Richard Cartwright Austin emphasizes:

> The sense of beauty is not a luxury. It is a matter of life and death for us and for the world. We have an obligation, not simply to ponder the meaning of beauty, but to open ourselves to experience beauty and to follow where beauty draws us. This distinctive human vocation can help us develop an ethical relationship with the natural world.[9]

Balancing Reason and Emotion

Reason alone cannot make people care, as feeling alone cannot guarantee that people will respond rationally and effectively. A balance is needed between reason and emotion before there can be a complete and effective transformation of self-love into other-love. And it is then that love, as concern for all living beings and the environment, becomes the supreme ethic.

Psychologist Erich Fromm has identified four elements of mature love: *care*, the active concern for life and growth; *responsibility*, the desire to respond to others' needs; *respect*, to look at and to recognize others' uniqueness; *knowledge*, combining impartiality with participation and intimate identification.[10] Attorney-humanitarian and farmer Robert F. Welborn sees compassion and reason as vital faculties, and contends that "compassion is the means for spiritual oneness, for the unity of all life and beauty . . . Reason gives us the understanding that we achieve happiness only with the life and beauty of Nature. It gives us that capacity to preserve and enhance that life and beauty." (personal communication)

Separation and imbalance between reason and emotion are the causes of much harm to person and planet alike. Too much emotion can lead to irrationality, impulsivity, and even emotional disease. Too much reason can lead to cold, unfeeling rationalism, excessive control, and even spiritual death. The emotional disconnectedness of intellectual rationalism is a disease of the times. Consciousness entails more than being rational: it also involves feeling and that deeper intuitive wisdom for which instrumental knowledge is no substitute. Without compassion and wisdom, such knowledge becomes linked with power, and its consequences which history documents have been spiritually and ecologically detrimental to humanity. Dostoevsky wrote in *The Brothers Karamazov*, "Love God's creation, love every atom of it separately, and love it also as a whole; love every green leaf, every ray of God's light; love the animals and the plants and love every inanimate object. If you come to love all things, you will perceive God's mystery inherent in all things; once you have perceived it, you will understand it better and better every day. And finally you will love the whole world with a total, universal love.

Love the animals: God has given them the beginnings of thought and untroubled joy. So do not disturb their joy, do not torment them, do not deprive them of their well-being, do not work against God's intent. Man, do not pride yourself on your superiority to the animals, for they are without sin, while you, with all your greatness, you defile the earth wherever you appear and leave an ignoble trail behind you—and that is true, alas, for almost every one of us!"[11]

Beyond Morality

Children should not be disciplined and indoctrinated by parents and teachers with moral codes that simply preach right from wrong. Such morality is a poor substitute for love and for the respect and concern for others that arises spontaneously when there is empathy. They should be encouraged to empathize "even with the least of these." Where there is empathy and compassion, there is no need for moral codes. Another of Lao Tzu's statements emphasizes that the essence of humane education is compassion and respect, not morality. "When Tao (all-embracing reverence) is lost, virtue follows. When virtue is lost, benevolence follows. When benevolence is lost, righteousness follows. When righteousness is lost ritual follows."[12]

Imposing a rigid morality upon children as to how they should treat animals, and each other, will impair their emotional development and personal freedom as adults, in whom freedom is a function of three attributes: humility, responsibility, and response-ability. As Lao Tzu observed: "Surrender yourself humbly: then you can be trusted to care for all things. Love the world as your own self; then you can truly care for all things." He also advised, "Give up sainthood, renounce wisdom (cleverness) and it will be a hundred times better for everyone. Give up kindness, renounce morality, and men will rediscover filial piety and love."[13]

When morality is based upon rationalism and not upon feeling or empathy, and when passion arises through the prisms of reason and not from the heart's instinct, there is great pretense and contagious ignorance. There is neither humility nor wisdom, only empty arrogance and the hollow meaninglessness of existence. Where there was once the benevolence of spirit, now there is the patronage of superiority and guilt. Where there was once love, now there is fear, for without love there can be neither trust nor respect. Now it seems we neither trust life nor respect death. And so we also no longer trust each other or respect other *inferior* living beings, animals, plants, rocks, rivers, and all. We even call them *things*. We fail to perceive some living entities, like rivers and mountains, as alive. But they are complex beings that are like ourselves in myriad ways.

Today, rational morality leads us to protect, but not cherish these *things*, these numinous existences of nature, for reasons primarily of utility. But we should look more closely at the other possible sources of response—aesthetic, humanitarian, and egalitarian. It is in and through these that we go beyond rational morality and the consensus legality of the times. Focusing on these leads us back to feelings and clarifies our perception. Through this we can begin to understand why Lao Tzu saw morality and legality as arising from the absence of empathetic sensitivity and ethical sensibility, and as self-limiting substitutes for these other virtuous gifts of grace. He advised us to beware of those who saw law and order, loyal obedience, and punishment in the name of justice as absolute values that must be upheld and defended without question. He was that kind of spiritual anarchist, so rare today, who was inspired by love's compassion, and not by fear or anger's frenzy, to help heal the world by restoring the Tao—our relationships with all. And this healing and restoration, as he warned, can never come simply through law and order, rational morality, and the suppression of our instinctual and emotional authenticity.

Since 1963 when Konrad Lorenz published *On Aggression* in which he proposed an innate (i.e. genetic) basis for human aggression,[14] there has been much debate over whether our behavior is influenced by nature (our genes) or nurture (how we are raised). Could there be a gene, or innate basis for empathy and altruism? More recent behavioral studies have shown how nature and nurture interact during critical stages of social development against the background of generations of natural selection for certain traits like aggression and altruism, territorial defense and maternal behavior, and coping with social conflict and stress.

Frans de Waal traced the origins of empathy, altruism, and evolution of morality in nonhuman animals.[15] Aside from the fact that humans are not the only beings on earth endowed with these "higher" attributes, comparative studies of human and nonhuman behavior help give us insights about our own nature, especially with regard to the natural selection of those innate genetically transmitted capacities for human aggression and selfishness, altruism and compassion. We humans, like our closest primate relatives that de Waal studied, are chimeric beings with innate potentialities of both good and evil. That science has proven animals can act as moral agents and show compassion and empathy should help encourage public acceptance of animal rights. Acceptance of the dark side of human nature, of what Van Rensselaer Potter calls our "fatal flaw,"[16] should encourage parents, teachers and social reformers to be alert to its manifestation in children and help educate children to understand and therefore gain better self-control of this dark side. In the process, their empathic and ethical sensibilities will be awakened and directed toward bioethical awareness and behavior. Potter describes our fatal flaw "as the tendency to emphasize the here and now and not to worry about the long-term future." It is the desire for immediate gratification, epitomized by consumerism. He sees this flaw as a major limiting factor in the blossoming of global bioethics. If Frans de Waal's apes and monkeys can teach their offspring self-control and concern for others, then surely we can do no less with ours.

Changing Human Behavior

Attitudes of indifference—why bother being ethical and caring in this competitive, dog-eat-dog world where so many are caught in the mayhem of self-interest—are self-limiting and only add to the tragedy of the human condition. What goes around, comes around, so it is enlightened self-interest to avoid bad karma. When we harm the earth, as eco-philosopher Henry K. Skolimowski has elegantly shown, we violate the "eleventh commandment" of ecology and we then harm ourselves.[17] Similarly when we violate the "twelfth commandment" of not treating animals as we would have them treat us, cruelty toward animals can lead to violence toward people.[18]

The ruthless tycoons of commerce and the industrial world are no different from some impoverished and disenfranchised third-world peasants who will also lie, cheat, steal, and even kill to secure their needs. Van Rensselaer Potter has proposed that this potentially fatal flaw in human nature is genetic in origin.[19] He contends, "The fatal flaw is probably in all of us in our 'animal' genes, but is countered by a set of 'ethical' genes that moderate the 'flaw' with a concern for future survival." This flaw is as biochemical and neuro-hormonal as it is cognitive and behavioral in its manifestation. The trait of extreme selfishness is, according to Potter, an ancient survival program. It translates into "I-Me-More," not "I-Thou-Share," limiting cognitive/affective/intellectual reflection from any perspective other than self-interest. If aroused by others' suffering, the trait limits our empathy and may even derive vicarious pleasure from other's demise. The flaw is evident in the rationalizations, indifference, and denial of the harmful consequences of those individuals caught in this immature, egotistical state of mind. The flaw needs to be recognized by child psychologists, educators, and others whose challenge lies in helping children control this inherited survival program of "I-Me-More" and in awakening the more altruistic genetic program of "I-Thou-Share."

It is doubtful that this trait can be eliminated. If it is indeed an inherent aspect of human nature, that would mean eliminating the entire species, which would be unethical, even if some believe it would be ecologically beneficial. Other solutions are to control the trait with genetically engineered and conventional pharmaceutical products like injections, pills, and implants. But these try to control the wrong things—anxiety, depression, mania, violence, impotence, infertility, and other stress-related disorders including early-onset senility—which are in part the symptomatic reactions to and consequences of this genetic defect. Since this potentially fatal flaw in human nature cannot be eliminated by genetic engineering, or other interventions, it is best controlled by individuals and by cultural evolution. In many preindustrial societies, this was instigated usually in early adolescence by various rites of initiation and traditional teachings meant to keep the "I-Me-More" program from dominating the psyche.[20]

Ethics and the Recovery of Our Humanity

The recovery of self from civilization is as difficult as the discovery of self in nature, because what is left of the natural world and the seeds of our own humanity in the collective consciousness of society have been so diminished by human ignorance, arrogance, and spiritual corruption. But until our humanity is recovered, the realist, the pessimist and the fatalist are one; it is realistic to believe that if we do not become more human, i.e., humane, one can only be pessimistic about the fate of our humanity. Through compassionate action and through bringing bioethics to life, the seeds of our humanity may be saved. This in part entails redefining what it means to be human. Then the evil consequences of short-sighted egotistical and anthropocentric actions and beliefs, that are so often touted as being progressive and for the greater good, may be averted. This means, for example, public opposition to genetic piracy by the life science industry that is patenting and then genetically engineering the seeds of crops developed over millennia of good science and bioregional social and economic cooperation by sustainable, indigenous farmers in the Third World.

Transpersonal psychologist Abraham Maslow found optimism through his research into the transformative effects of "peak" experience on people, rich and poor.[21] He noted that preexistential psychologists like William James regarded these as "varieties of religious experience," while the phenomenologist Merleau-Ponty linked this psychological state of self-realization and divine revelation with the "beingness" of Zen Buddhism. In Hinduism (Vedanta), Maslow's peak experience is called self-realization: *Tat-tvam-asi*, meaning "thou are that," the sudden apprehension of the oneness and sacred unity of all life.

I do not share the optimism of Abraham Maslow, and of humanistic psychology, that when our basic needs are met, higher values and aspirations toward altruism and self-realization arise spontaneously. Transpersonal psychology recognizes the transformative influences of parents, society and education in enabling a child to transcend egotism and the self-limiting horizon of seeking immediate need-gratification. But the child's development is arrested and impaired when the dominant culture does not facilitate development by fostering empathic sensitivity toward others and bioethical sensibility, ideally enhanced by a metaphysical or religious perception of the sacred unity and interdependence of all life. Such an arrested development seems to be the case in our highly competitive, individualistic, and materialistic industrial-consumer society. Transformative "peak" experiences are not easily realized, since the contemporary values of the dominant culture offer a plethora of consumables and distractions that impoverish, confuse, and sicken, rather than nurture and inspire the soul. The status quo and world view of consumerism and industrialism is maintained through the educational process, through media manipulation of truth, and control of information, along with desensitization and denial. But for how long, we must ask, can this status quo be maintained, the harmful consequences of which must be seen as evil, because it is neither just nor sustainable. Unless the bioethical seeds of our humanity, along with the seeds of biocultural diversity and sustainable agriculture, are protected from extinction, this worldview will be our nemesis.

We will know that our ethical sensibility has been restored when, as humanitarians, we feel the suffering of other animals as our own. Christians might empathize with the Earth as Christ crucified. As consumers, we might realize the moral imperative of eating with conscience. And as parents and responsible citizens we might teach compassion, ahimsa, and reverential respect for all life through example. This is the first step that later chapters will detail, for the healing of culture and agriculture, for the practice of a more effective medicine, and for the foundation of a sustainable social economy. It is also a vital step for all of us to begin to heal ourselves, each other and all our relationships, as we bring balance and harmony to our families and communities, and to the pursuits of commerce and industry. But so long as there is even one dog starving at our gates, and no wolf singing in our forests, we and the world may never be well. Compassion is a boundless ethic and is the leading ethical principle that is a universal absolute.

We all need to ask ourselves what the purpose is of human existence other than the continuation of the species. What are we living for? To multiply and consume the earth? To improve nature through genetic engineering? Bioethics calls for us each to complete our personal inventories of how much we take and harm and how much we give and heal. How do we as a community or species compare to the ways of the wolf and the wisdom and benevolence of the trees? In posing these deeper questions that the more conventional anthropocentric schools of ethics have not addressed, bioethics helps us to examine why our humanity is as endangered today as the last of the wild, and in finding the connections, find solutions. When we harm the Earth, we harm ourselves. When we demean and violate the sanctity of life, poisoning the meadowlarks and the columbines, we demean our own humanity and divinity within.

If we are to save our humanity and the life and beauty of the Earth, we must obey what native Americans call our original instructions. In the Judeo-Christian tradition, this means obeying the Golden Rule. We are to be the custodians of God's Creation, called to dress and to keep the Garden of Eden, to cherish, revere and protect all creatures and Creation. In practical terms what is called for is planetary CPR—conservation, preservation, and restoration. This is not possible so long as society sees progress as industrial growth and the proliferation of consumer wants, goods and services that create jobs but rob future generations by squandering nonrenewable natural resources and polluting the environment. Industrialism and consumerism will never help the millions of malnourished people in resource-depleted and overpopulated regions of the world. Industrialism and consumerism mean over-consumption and a temporary rise in the standard of living for those fortunate few who can get jobs. When we factor in the number of people living today, there can be no tomorrow for western industrialism because it is nonsustainable. It is socially unjust, ecocidal, and genocidal. What is called for is a redefinition of what it means to be human, in terms of our responsibilities for the life, beauty, and future of the Earth. This entails a redefining of the term "progress" and a radical shift and transformation of the global economy based on the bioethical principles of sustainability, frugality, and social justice, including transgenerational equity and ecojustice. We cannot have the earth and eat it too.

Confucius recognized some 3,000 years ago the social and spiritual significance of frugality—eradicating the desire for things. He advised in his treatise *The Great Learning* that:

"Among the ancients, he who wished to have the shining virtue illuminated throughout the world, first governed his nation well. Wishing to govern his nation well, he first managed his family in good order. Wishing to manage his family in good order, he first cultivated his person. Wishing to cultivate his person, he first rectified his heart. Wishing to rectify his heart, he first rendered his thoughts sincere. Wishing to render his thoughts sincere, he first let his innate intellect reveal itself. The way to reveal innate intellect is to eradicate the desire for things."[22]

Saving the Earth and our humanity entails liberation from human selfishness, ignorance, fear, arrogance, hatred and greed. This task for humanitarians, ethicists, and others today is daunting indeed. The spiritual evolution of our species is a developmental, birthing process, and we are our own midwives.

The love and respect that we give and receive, not what we own, determine the spiritual quality of our lives, but not until we have examined what we live for and what we would die for. As an old Serbian saying reminds us, "We should be noble because we are part stardust, and we should be humble because we are part manure." Until we accept death without fear, we will pit life against death and life against life.

To be spiritual is to be ethical and in tune with the sacred, through feeling, faith and reason, and through right action and relationship. Another view of spirituality is to see it simply as being open and responsive to the sacred, to the intimate revelations of divinity in the beauty and mystery, tragedy and suffering, joy and wonder of life. Sustainable living therefore entails living in the spiritual as well as the material by bringing bioethics to life. The sacred is that which moves us toward grace and the realization of divinity in the awakening of our sense of wonder, mystery, gratitude, reverence, and celebration.

Spirituality and Biospiritual Ethics

The term spiritual, like the word ethical, can be used as an adjective—as a descriptor of our behavior, of a kind of experience and of our state of being or awareness. It is founded upon the realization of the essential unity and interdependence of all life which engenders a transfiguring passion and loving concern for all sentient beings. From this passion the universal and universalizing ethic of compassion is born, which we endeavor to incorporate in all our doing and with all our being. From this metaphysical perspective, bioethics is derivative of the realization of the sacred, and leads to a more spiritual way of life. Hence, my earlier use of the term *biospiritual* ethics.[23] It is indeed encouraging that Van Rensselaer Potter's secular bioethics (discussed in Chapter 2), in its non-reductionist, non-anthropocentric "global" framework, is now being adopted by Catholic and other theologians to serve as a bridge between religion and metaphysics, and concern for the natural world—God's Creation. This is precisely the kind of integration that I have long wished for under the banner of biospiritual ethics. But concern for the natural environment and for biodiversity will remain essentially anthropocentric, albeit enlightened, until we accept that animals and all sentient beings are not only morally considerable, but worthy of equal consideration. Divine presence and consciousness of the sacred are one and the same. Bioethics and the biospiritual principle of reverential respect for all life enable us to engage and enjoy this essential unity: communion with the Earth community, and with the life universal of an emergent cosmos of divine conception, revelation, and redemptive self-realization.

CHAPTER 2

Bioethics:
Its Scope and Purpose

Van Rensselaer Potter sees the term bioethics as the "science of survival" in the ecological sense, an interdisciplinary approach to ensuring the preservation of the biosphere through Earthcare, People-care, Animalcare, and Healthcare. My own definition of bioethics is simply ethics for life and for living. Bioethics helps us choose how best to live in peace and harmony within the broader biotic, or life, community for our own good and for the good of other members of this community. Bioethics extends the Golden Rule to all living beings.

Bahai Faith: He should not wish for others that which he doth not wish for himself, not promise that which he doth not fulfill." *Gleanings*

Buddhism: "Hurt not others in ways that you yourself would find hurtful." *Udana-Varqa*, 5.18

Christianity: "As you would that men should do to you, do ye also to them likewise." Luke 6:31, *King James Version*

Confucianism: "Surely it is the maxim of loving-kindness: Do not unto others that you would not have them do unto you." *Analects*, XV.23

Hinduism: "This is the sum of all true righteousness: deal with others as thou wouldst thyself be dealt by. Do nothing to thy neighbour which thou wouldst not have him do to thee after." *The Mahabharata*

Jainism: "A man should treat all creatures as he
                       himself would be treated." *Sutrakritainga*,
                       1.11.33

Judaism: "What is hateful to you, do not to your fel-
                       low man. That is the entire Law, all the rest
                       is commentary." *The Talmud*, Shabbat, 31a

Islam: "No one of you is a believer until he de-
                       sires for his brother that which he desires
                       for himself." *Sunnah*

Taoism: "The good man ought to pity the malig-
                       nant tendencies of others; to regard their
                       gains as if they were his own, and their
                       losses in the same way." *The Thai-Shang*

Zoroastrianism: That nature only is good when it shall not
                       do unto another whatever is not good for
                       its own self." *Dadistan-iDinik*, 94.5

The dictum, what is good for the Earth is good for us, entails putting bioethics into action in the many ways detailed in subsequent chapters. Passive reverence for life alone will not suffice. The more we accept that it is the moral obligation or sacred duty of all of us to give equal and fair consideration to every sentient being and to act accordingly, a "golden mean" will begin to be established, as our own good parallels and complements the good of the life community. Utopian and unattainable as this ideal may seem of what essentially amounts to a new covenant of planetary care for enlightened humanity, I am optimistic.

The recent plethora of books addressing the subject of bioethics is a positive sign of the increasing recognition of the relevance of ethics to human affairs.[1] This is a long overdue corrective because the subject and discipline of ethics has become highly institutionalized, increasingly anthropocentric and divorced from the very fabric of our daily lives. There has been a notable lack of linkage until recently between theology, spirituality and ethics concerning our duties and responsibilities toward the nonhuman creation. Ethics has also been supplanted by scientific determinism (scientism) and co-opted, perverted, and then reified by vested interests seeking ways to rationalize the continuation of activities that cause great harm to animals, to the environment, and to the consumer populace. Medical and veterinary ethics, for example, have had more to do with professional conduct and the avoidance of legal suits than with social, environmental, and economic factors in human and animal suffering and disease. But with the recent incorporation of bioethics, these ethical issues are being more fully addressed by both the medical and veterinary profession.

Bioethics is a term that is becoming widely used today. It entails the objective appraisal of how our values, desires, and actions affect others, including animals and the environment. We have medical bioethics, which focuses on such ethical issues as euthanasia, surrogate parenting, and genetic engineering involving human beings. Extending bioethics to the evaluation of laboratory animal care and use will do much to improve the well-being of animals, and it will make biomedical research publicly more acceptable and medically more relevant. Weighing the costs of animal life and suffering against the potential benefits of developing new medical, surgical, and diagnostic procedures is too narrow an evaluation. Without a broader bioethical evaluation, alternatives to using animals, and alternative procedures to alleviate and identify the causes of human sickness and suffering will not be forthcoming. These and other bioethical issues have been deliberated by the World Council of Churches, among other groups. Increasingly, these groups are beginning to apply bioethics to a host of other social and environmental issues.

It has been argued that animals can't have rights because they lack the capacity to be moral agents. Likewise, trees and natural ecosystems cannot have legal standing because they do not have manifest interests or entitlements, while a corporation can have legal standing.[2] Our laws and our ethics, like our economy and industries, are clearly too narrow in scope; they do not include animals, plants, and the entire biotic community of the Earth's vast ecosystem as being morally or otherwise considerable, except as property and resources. Bioethics, as defined by Van Rensselaer Potter,[3] expands the scope of our moral obligations and spiritual awareness to embrace the entire biotic community. Father Thomas Berry calls the natural sentient world the "life" community (which he sees as a community of subjects rather than a collection of objects), of which we are but one of many citizens.[4]

Bioethics also incorporates *biosophy*, the wisdom derived from a scientific (especially biological and ecological) and empathic understanding of life. Biosophy is reinforced by what Harvard biologist E. O. Wilson[5] calls *biophilia*, our innate sympathetic affinity for the life and beauty of the natural world, which is the spiritual ethos of the "deep" ecology and animal rights movements.[6] This incorporation makes for what Potter calls *global bioethics*,[7] that serves to bridge the human and the nonhuman realms of the life community as it links also all human spheres of activity (see Fig 2.1) with concern and responsibility for the life and beauty of the earth.

We need to consider how animals, plants, and ecosystems can be harmed by various human activities. The ethics of compassion and reverence for life move us to avoid harming other sentient beings. Ethical behavior requires great vigilance because of our innumerable habitual "sins" of omission and commission that we commit—for example, as consumers of factory-farmed animal produce, as users of electricity and gasoline, and as users of various medicines that are tested on animals and that we inevitably excrete into the environment, causing further harm to natural ecosystems and to ourselves.[8]

For those who would give a political label to global bioethics I would say that it is neither left nor right, but both. It is holistic and anarchistic in the sense of Peter Kropotkin's view that there is no strict hierarchy in nature.[9] Nature is an anarchy, based on mutual cooperation and mutually enhancing co-evolved symbiotic relations. Ecosystems function as a whole without a ruling class. Predators are merely at the apex of the food chain. They have as much "rule" as the mycorrhyza in the soil. So I would say that "deep" and global bioethics is the rational foundation for a radical, spiritual anarchy; in essence, participating in the democratic "holarchy" or circle of life wherein the redefined, reoriented, and reintegrated human species continues to consume and use the earth's resources to sustain its existence, but with less harm, and works to restore, protect, and cherish the life and beauty of the Earth.

Putting reverence for life into action is the essence of a transspecies democracy and is the "heretical" anti-establishment view of the spiritual anarchist who gives all sentient beings equal consideration. E. L. Ericson contends that "such [a] society of 'heretics' is an uncomfortable and unstable community, yet that is what true spiritual democracy represents." Ericson believes that the free mind creates the free society, not the other way around.[10] In the absence of this democratic worldview, the deterioration of the environment will continue, accelerating incrementally as the human condition also deteriorates, at the individual level and also at the levels of community and the global economy.[11]

Global bioethics is no panacea for the problem of this and future ages, but it offers a compass of reason and compassion for us to find the way to security and fulfillment without causing so much harm that we continue, from age to age, to bring evil into the world and

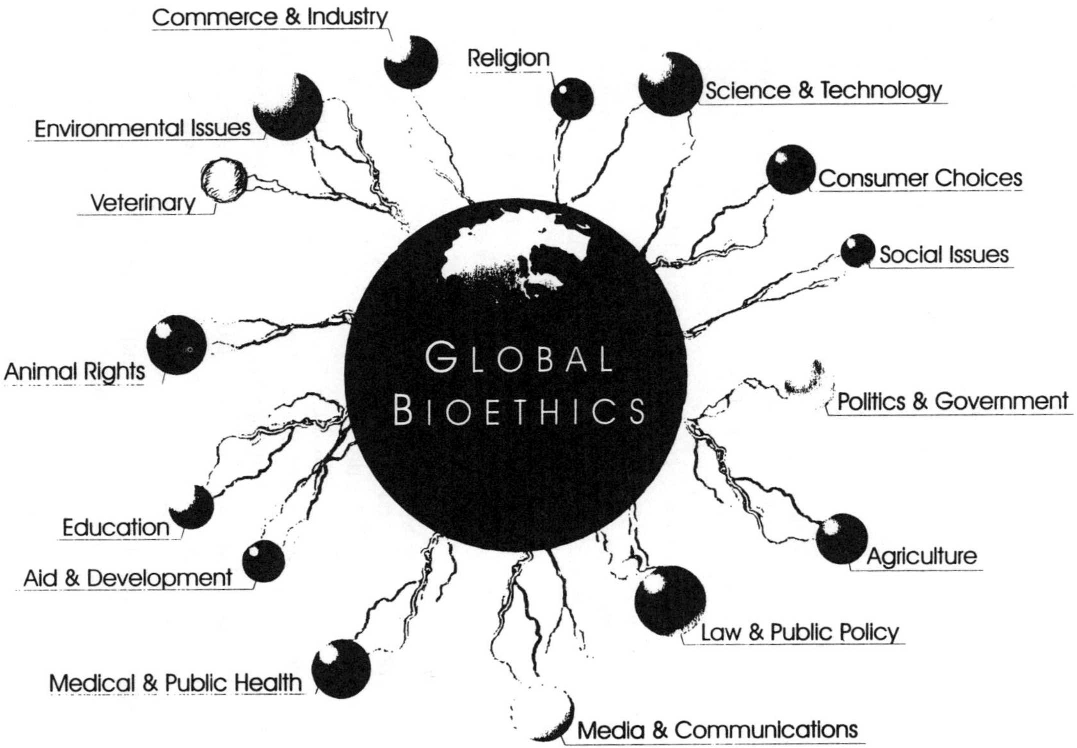
Commerce & Industry
Religion
Science & Technology
Environmental Issues
Consumer Choices
Veterinary
Social Issues
Animal Rights
GLOBAL
BIOETHICS
Politics & Government
Education
Agriculture
Aid & Development
Law & Public Policy
Medical & Public Health
Media & Communications

Fig. 2.1. Schematic representation of the bridging function of global bioethics linking various realms of human activity and issues of concern, including our treatment of animals and the environment.

harm ourselves in the process. Reflecting on the emergence of the earth's most recent mammalian creation, *Homo sapiens*, biologist Loren Eisely wrote, "The Eden of the eternal present that the animal world had known for ages was shattered at last. Through the human mind, time and darkness, good and evil, would enter and possess the world."[12]

It is my hope that applied global bioethics will help dispossess the world of this evil and heal the chimeric, schizoid condition of the half-formed, unintegrated human mind, by shattering the artificial dualities of distorted and confused perception that lead us to believe we are somehow separate from God and nature, and superior over animals and the rest of Creation. In this state of mind we can quickly "split," severing our empathic connections and becoming in-human. Global ethics is the triune bridge between people, animals and nature (PAN), providing a rational basis for an earth- or Creation-centered spirituality, politics and economy. To be human is to be part of nature; human, humus (soil) and humility are etymo-logically linked. To act apart from nature, treating other sentient be-ings without compassion or equal and just consideration, is to be inhumane—therefore, inhuman.

In a recent essay Potter shares Herman Daly's view that sus-tainable growth is an oxymoron and insists that "the idea of sus-tainable development should be abandoned and replaced by the idea of sustainable survival."[13] Potter goes on to observe that "enlight-ened anthropocentrism calls for control of human fertility and sees the human species in the context of the total biosphere. . . . We need a global bioethic that will guide good intentions and harness the will to power. Global bioethics calls for good intentions that are covered by five realistic virtues: humility, responsibility, interdisciplinary competence, intercultural competence, and compassion."[14] Other vi-sionaries like Paul Kennedy and Hans Küng likewise see that global survival is more dependent upon the international acceptance of a global bioethics than on global economics.[15] As Potter notes, "The legacy of Adam Smith, the invisible hand and the right to life, liberty and the pursuit of happiness must be tempered by global ethics."[16]

Bioethical Principles and Virtues

We need to recognize the vital role that bioethics can play in helping students develop moral integrity and ethical sensibility and responsibility. But the role of bioethics in facilitating normal development both emotionally and intellectually cannot be effective in an institutionalized or cloistered environment while the outside world of human affairs is not governed by the same bioethical principles.

Ethics is something far more profound than a subject of academic study and scholarly discourse. It does not mean simply learning how to choose between good and evil and discriminate right from wrong. Ethics helps us go beyond such dualities, and as we incorporate the virtues and principles of bioethics we become ethical beings. In that more spiritual state of being-in-awareness, through reason and empathy we come to that point of self-realization where the Golden Rule becomes second nature: one's own selfhood is no less and no more sacred than that of any other sentient being. In essence, therefore, ethics is part of the spiritual alchemy of the human psyche. It has more to do with the rediscovery of what it means to be alive and to be human, with responsibilities and longings, and with virtually unexplored horizons of creativity, joy and wonder, than with a set of rules and codes of conduct that we should obey without question for our own good and for the good of society. The basic principles of bioethics, grouped into three arbitrary categories (activist, reformist, and transformist—see Fig. 2.2) unify our hearts and minds, our feeling and thinking, enabling us to visualize and actualize our humanity, the essential goodness inherent in human nature, through ethical behavior.[17] The possibility of creating a more enlightened age and a humane society may then become a reality for the next generation to pursue with vision, passion and commitment.

Bioethics essentially reincorporates the virtues into our lives. It is the heart of reason's response to a fundamentally spiritual crisis that is integral to correcting the global environmental and socioeconomic problems—the unintended calamities of materialism, industrialism and colonialism—that science and technology alone cannot solve. Bioethics speaks to our moral ecology and imagination, its basic principles being the virtues that can guide and inspire us toward health and wholeness, and toward world peace, justice, fulfillment and securing the integrity of Creation.[18]

The basic bioethical principles (see Fig. 2.2) are the seeds of our humanity that must be recovered now, and then sown and harvested more and more by each generation to come, transforming the human species from a destructive planetary infestation into a protector and reverential participant in enhancing and celebrating the life and beauty of the Earth.

A healthy humanity is concerned about its humanity—how compassionately it acts toward its own kind and toward other sentient

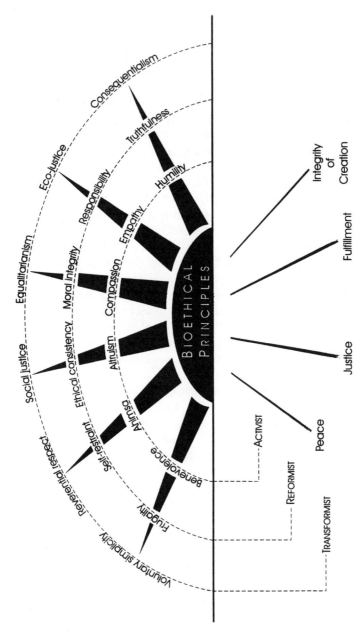

Fig. 2.2. Activist, reformist, and transformist principles and virtues of bioethics form complementary rays that illume the paths toward peace, justice, fulfillment and preserving the integrity of Creation.

beings and the earth itself. It has respect for all life because it realizes that when it damages the environment, it harms itself. Bioethics, in this regard, is a field of self-investigation and enlightened self-interest. It also provides a foundation to establish meaning in our lives.

Bioethics offers a holistic, rational appraisal of our place in the world and suggests how best we can live for the good of the biocommunity of the planet. It mandates that equal and fair consideration be given to human rights issues, animal rights issues, and environmental concerns. It includes a temporal principal of transgenerational equity—concern about the well-being of future generations and respectful understanding of the wisdom and folly of our ancestors. We should neither forget our history, lest we repeat it, nor forget that "we do not own the land, we borrow it from our children."

Bioethics can be an antidote to the prevailing dominionistic attitude toward life. The subjugation of minorities and other life communities will continue as will war and other forms of violence until we reverse the belief that we are superior and apart from nature. The polemicized rhetoric and bickering within and among frustrated factions of the human, animal, and environmental rights movements are reconciled by the integrative approach that applied bioethics provides.

Within what some call the *establishment*—meaning the government-industrial complex—bioethics is also taking root. Ethical conduct, ethical advertising, ethical products, and full cost-accounting are beginning to appear on its agenda. Protection of endangered species, humane treatment of domestic animals, sustainable use of agricultural and other natural resources, loss of biodiversity, global warming, air pollution, national economic security, and industrial-economic sustainability—these and other issues fall within the sphere of bioethics.

Bioethics can help lead policy makers and corporations toward a more holistic approach to clarifying and finding the most equitable and ethical means to achieve their desired ends. Ultimately, bioethics helps establish a common ground for all the different values and desires we all have. Different voices are heard—those who speak for the sick and dying, those who speak for laboratory animals, those who speak for spotted owls and sacred forests, as well as those who speak for native peoples and cultural, as well as biological, diversity. The democratic process is facilitated by the principle of giving equal and fair consideration to all sides or aspects of a given issue concerning human, animal, or environmental rights. Bioethics is clearly a philosophical integration of human, animal, and environmental rights. It fosters an Earth- or Creation-centered worldview, a paradigm shift—what E. F. Schumacher, British economist and father of ecological economics (eco-nomics), termed *metanoia*.

Decisions and full-cost accountings based upon bioethics should include scientific, economic, legal, moral, social, environmental, and compassionate considerations. They are not purely "science based" as is the trend today in such matters as wildlife protection, habitat conservation, FDA approval of genetically engineered bovine growth hormone (rBGH), and USDA approval of the release of genetically-engineered organisms into the environment. The European Community (EC) ban on injecting cows with rBGH will cost the manufacturers and investors in this product billions of dollars. The EC parliament ruling was based primarily upon the bioethical principles outlined in this book, as was their decision to put a moratorium on the patenting of genetically engineered animals. In light of these recent developments, it would be enlightened self-interest for corporations to incorporate bioethical criteria into their research and development decision-making processes. Developing new products like rBGH in an ethical vacuum is good for neither stockholders nor the corporate image. The moral complexity of many contemporary issues, notably in the area of genetic engineering biotechnology, is considerable and cannot be subsumed by a "business as usual" attitude.[20]

The moral component of bioethics plays a central role. It is based on the principle of ahimsa, of avoiding harm and injury in the process of furthering human interests and the good of society. Whatever the issue, bioethics begins with the premise that all living beings and natural processes have purpose, if not interests. The intrinsic value of other sentient beings and the inherent worth each being has in relation to its community are therefore acknowledged as deserving moral consideration.

The instrumental or extrinsic value of a given life form may appear insignificant when we make our value judgement of its entire being on the basis of its degree of sentience or intelligence. Such judgement is invariably wrong. Consider that without lowly fungi in the soil, our crops and forests would be sickly, grow poorly, and we would suffer the consequences.[21] We should, therefore, be mindful "of the least of these," and not continue to destroy them directly with agricultural chemicals and indirectly with industrial pollution, most notably acid rain.

Every community—human and nonhuman—also has intrinsic value, not only value to its members (in terms of security, continuation, etc.) but also to the larger life community of the planet's homeostatic and regenerative biospheric ecosystem. As Aldo Leopold wrote in his seminal book, *A Sand County Almanac*, "A thing is right when it tends to preserve the integrity, stability and beauty of the biotic community. It is wrong when it tends otherwise."[22] And as E. O. Wilson suggests, "The more we know other forms of life, the more we enjoy and respect ourselves. Humanity is exalted not because we are so far above other living creatures, but because knowing them well elevates the very concept of life."[23]

Actions and products that disrupt others' lives or their life processes, which make up our life-support system, cannot be accurately forecast by the scientific method alone. When our thinking is linear, we focus too narrowly on the goal or gain, not broadly enough in terms of what means we utilize and to what ends. Nor do we consider what the long-term environmental, social, and other consequences, risks and benefits, might be. Bioethics demands that we pay attention to these various means and consequences, and how our means and ends might violate the principle of ahimsa. We are, for the sake of humanity—our dignity and integrity—bound to avoid causing harm or injury to any sentient being or to the biospheric ecosystem, when such harm or injury can be avoided. The all encompassing scope of bioethics allows for a higher degree of risk predictability. This is because bioethics operates from the moral principle of respect for all life. This means we must all strive to live nonviolently, because when we directly or indirectly harm or injure sentient beings or the environment, we not only ultimately harm ourselves but demean and impoverish ourselves and the earth in the process. [See Chapter XII for discussion on non-injury and nonviolence (ahimsa).]

Bioethics posits that all life has been created by forces we do not yet fully comprehend and that life is only ours in sacred trust. One of the founders of bioethics, Albert Schweitzer, wrote, "Ethics is in its unqualified form, extended responsibility with regard to everything that has life." He is unequivocal about the sense of duty that bioethics instills, stating that "the universal ethic of reverence for life shows the sympathy with animals, which is so often represented as sentimentality, to be a duty that no man can escape."[24] In a highly pragmatic sense, bioethics teaches us that when we take care of the earth, the earth will feed us. It also teaches that when we live in harmony with nature, nature will take care of us.

In conclusion, bioethics provides the holistic framework to help us deal more effectively with a host of issues that we face in our personal and professional lives. It enhances dialogue and facilitates conflict resolution, and because of its democratic process, it provides a firm foundation for a just and humane society. As later chapters will demonstrate, the incorporation of bioethics into our public, private, and corporate lives is enlightened self-interest. There is a sector of the public that is still opposed to environmental protection and animal rights and sees such concerns as actually against the public interest. In the next two chapters, this conflict will be investigated, as well as the tensions and conflicts between rights and environmental ethics, and solutions will be sought. The following synopsis of global bioethics shows the scope and relevance of the discipline of enlightened self-interest.

Synopsis:
Global Bioethics: Bringing Life to Ethics

- Global bioethics calls us to give equally fair consideration to three spheres of moral concern:

  > human well-being (rights and interests),
  > nonhuman well-being (rights and interests), and
  > environmental well-being (biodiversity and ecosystemic integrity).

- Global bioethics calls us to be accountable for our actions and appetites in relation to these three spheres; and to examine how well society, our politics, economies (industry and commerce), religious and other traditions, as well as our own personal lives, are in accord with the bioethical principles that unify these three spheres in the light and language of compassion, humility, and reverence for the sanctity of life.

- Global bioethics calls us to actualize our innate empathic sensitivity, moral sensibility and powers of reason, reflection, and self-control.

- Global bioethics calls us to consider the purpose and potential of human existence, the significance of the virtues that make us *humane* beings, and our duties and responsibilities for the Earth community, and for the integrity and future of Creation.

- Global bioethics calls us to understand and respect the cultural ecology of moral pluralism, and from this diversity of human beliefs, opinions, and desires, create a common ground of equalitarianism and respect for all life.

- Global bioethics calls us to develop a unity of spirit for more effective and immediate crisis management, conflict resolution, and humane intervention where the compass of compassion directs reason and action toward world peace, justice, environmental protection, and security and fulfillment for all sentient beings.

CHAPTER 3

Environmental Ethics
and Animal Rights:
An Overview

Eco-philosophers and environmental ethicists, especially in the US, have made various and significant contributions to the advancement of our understanding of our duties and obligations to the living biosphere—Nature—and why we ought to care and how. The first book to lay the foundation for environmental ethics was Leopold's 1949 *A Sand County Almanac*. His was the first clear voice to articulate the "land ethic" and to focus public attention on the need to manage, protect and respect the "integrity, stability and beauty of the biotic community."

In *A Sand County Almanac* Leopold wrote:

> This extension of ethics, so far studied only by philosophers, is actually a process in logical evolution . . .

> The first ethics dealt with the relation between individuals; the Mosaic Decalogue is an example. Later accretions dealt with the relation between the individual and society. The Golden Rule tries to integrate the individual to society; democracy to integrate social organization to the individual.

> There is as yet no ethic dealing with man's relation to land and to the animals and plants which grow upon it. Land, like Odysseus' slave-girls, is still property. The land-relation is still strictly economic, entailing privileges but not obligations.

> The extension of ethics to this third element in human environment is, if I read the evidence correctly, an evolutionary possibility and an ecological necessity. It is the third step in a sequence. . . .

> The land ethic simply enlarges the boundaries of the community to include soils, waters, plants, and animals, or collectively, the land.[1]

Echoing the sentiment of Leopold, and building on his germinal thesis, philosophers, ecologists, theologians and public policy makers subsequently helped develop an essentially American school of environmental ethics. Attorney Christopher D. Stone's examination of the question, Should trees have standing?, opened the legal and judicial systems to ethical scrutiny and environmental accountability.[2] Stone subsequently developed his ideas on the moral and legal status of animals, ecosystems and persons, rejecting the Monist concept of ethics (that, for example, may apply the same considerations in arguing for the protection of endangered species as we would in determining our duties toward our own kin) in favor of Moral Pluralism.[3] This is quite different from so-called situational ethics and moral relativity, since as Stone proposes, Moral Pluralism is necessary because of the diversity of ethical issues humanity is faced with today. The public good is now increasingly seen as derivative of the well-being of the forests, rivers, oceans, and entire biotic community of micro and macro-organisms that help create and maintain the atmosphere and climate and, through what biologist Lynn Margulis calls symbiogenesis,[4] maintain and modulate the frequencies and intensities of species and community transformations via biological selection, mutation, adaptation, and ecological succession, the sum of which is crudely referred to as "evolution."

The values of living beings and ecosystem communities have been the focus of considerable debate, ranging from the utilitarian perspective to the spiritual. There is growing recognition of the intrinsic or inherent value of these living entities, as well as of what Holmes Rolston III calls their "systemic value." He writes:

> "We might say that the system itself has intrinsic value; it is, after all, the womb of life. Yet again, the 'loose' system, though it has value *in* itself, does not seem to have any value *for* itself, as organisms do seem to have. It is not a value owner, though it is a value producer. It is not a value be-

holder; it is a value holder in the sense that it protects, conserves, elaborates value holders (organisms)."[5]

Rolston III responds to the allegation that we commit the "naturalistic fallacy" in finding value in nature, contending that "the danger is the other way round. We commit the subjectivist fallacy if we think that all values lie in subjective experience, and worse still, the anthropocentrist fallacy if we think that all values lie in human options and preferences."[6]

Philosopher Eugene C. Hargrove applied aesthetic value, arguing that as moral agents we have prima facie duties to protect and promote the beauty and integrity of naturally evolving ecosystems.[7] Yet others have adopted a "rights" approach, especially with reference to the moral right of living beings and ecosystems not to be harmed, some authors focusing on our duties and obligations toward animals, to protect endangered species, and their habitats.[8] A number of philosophers have sought to bring unity to the field of environmental ethics, notably T. Benton, Bryan Norton, and G. E. Varner.[9]

Arne Naess, to counter what he saw as a shallow, utilitarian and anthropocentric approach being taken by environmental ethicists and conservationists, developed the "deep ecology" philosophical movement based on the concept of biocentric egalitarianism (that holds all organisms as being equal).[10] Naess saw his advocacy of biocentrism as an antidote to the dualism and materialism of the modern worldview that regards humankind as being separate from nature, which is simply a resource for human use. Both "deep" and "shallow" ecology tend to give more priority to the biotic community as a whole, or to ecosystems, rather than to individuals.

Naess was critical of short-term technological fixes to environmental problems, which tended to create further problems. This "shallow" approach contrasted with the "deep ecology" approach which he advocated, that entailed a change in attitude and the development of an ecological consciousness. American environmentalists Bill DeVall and George Sessions popularized and elaborated on Naess' views in their book, writing:

> . . . deep ecology goes beyond the so-called factual scientific
> level to the level of self and Earth wisdom . . . [It] goes beyond
> a limited piecemeal shallow approach to environmental problems and attempts to articulate a comprehensive religious
> and philosophical worldview. The foundations of deep ecology

are the basic intuitions and experiencing of ourselves and Nature which comprise ecological consciousness. . . . Many of these questions are perennial philosophical and religious questions faced by humans in all cultures over the ages.[11]

The deep ecology movement, coupled with Edward Abbey's book *The Monkey Wrench Gang*,[12] helped give rise to the radical Earth First! Movement, judged by some as eco-terrorism and by others as appropriate civil disobedience against the rape of the earth. This ecological resistance movement is now worldwide[13] and is a prime example of bringing bioethics to life and putting ethics into action.

Criticism has been made of some mainstream environmental and wildlife protection organizations by Jim Davis and others who see larger organizations taking a conservative "win-win" approach, and making too many compromises because they fear losing political allies and corporate and private funding if they appear too "extreme."[14] Attorney Gary Francione similarly contends that the spirit of direction-action that characterizes the animal rights movement is being silenced by the "new welfarism" of some larger animal protection organizations that follow traditional welfare reform activities (investigations, education and legislation) and animal rescue operations under a facade of pro-animal liberationism.[15]

Direct, nonviolent action at the "grass-roots" level is still, however, very much a part of the environmental and animal rights movements.[16] Finnish ethicist Leena Vikka proposes that an ethical theory based on accepting that nature itself has intrinsic value could promote an attitude of respect for all life forms, including the human.[17] Such a naturocentric, rather than anthropocentric theory of values would, she believes, foster greater acceptance, that various nonhuman entities like animals, plants, and mountains, possess intrinsic value and therefore have intrinsic rights. The value of life requires, therefore, the right to live and flourish in each life form.

We should also add J. Baird Callicott's community value perspective (he now advocates ethical vegetarianism)[18] and Jay McDaniel's sacramental value.[19] When integrated, these different perspectives lead to a holistic view that takes us beyond anthropocentrism, affording a more biocentric worldview, like James Lovelock's Gaia hypothesis.[20] The view is that the Earth is a living, breathing life form in its totality of co-creativity, co-evolving life forms and natural systems and processes. From this perspective the whole planet becomes a guide for moral value.

I like R. Attfield's reasoning that it is the ability to flourish and develop that gives moral standing to living beings.[21] This is close to my own view that we need not debate how sentient or how much intrinsic, inherent, instrumental or extrinsic value living beings may have in order to accord them value. Rather, it is in beingness itself that all values arise from the miracle and mystery of the created world to which we are party, witness and participant. Values are secondary, if not anthropocentric, and derivative of that which is ineffable and is made profane by being reduced to some measure of human value and significance.

Spiritual Views

The biocentric view of many environmental ethicists has considerable resonance with the theocentric or spiritual orientation of other philosophers and theologians.[22] A biocentric or theocentric orientation helps overcome the limitations of anthropocentrism as evidenced in earlier writings addressing the question of human responsibilities toward the environment, like John Passmore's book *Man's Responsibility for Nature*.[23]

Henryk Skolimowski is one eco-philosopher who has taken a spiritual approach to environmental concerns, emphasizing the importance of Albert Schweitzer's principle of reverence for life, and the correlations between pollution of the physical, the mental, and the spiritual. He emphasizes, "What are ultimately responsible for the polluted planet are not dirty technologies but polluted minds—careless, unseeing, alienated. Contaminated minds produce contaminated environments."[24] Lutheran theologian Jurgen Moltmann contends, "If we humans are made in God's image—the *imago dei*—then human beings must love all of their fellow creatures with the Creator's love. If they do not, they are not the image of the Creator, and the lover of the living. They are his caricature."[25] An increasing number of theologians now interpret "sin" in the biblical sense as a refusal of humanity to accept its place in nature and to live conscious of its relationships with each other, animals, and the natural world.[26]

What seems to be emerging, possibly due to the influence of these and earlier theologians, is a reenchantment of science and nature by awakening a feeling for the supernatural or omnipresence of divinity.[27] This metaphysical worldview, which has considerable cross-cultural and interreligious resonance (and so cannot be quickly dismissed as cultural imperialism or ethical colonialism) was called the "perennial philosophy" by Aldous Huxley. He characterized it as "the metaphysic that recognizes a divine reality substantial to the world of things and lives and minds; the psychology that finds in the soul something similar to, or even identical with, divine Reality. The ethical that places man's final end in the immanent and transcendent ground of all being."[28]

Some critics of the spiritual/mystical orientation of deep ecology, like Murray Bookchin,[29] felt that it was too far removed from pressing economic and social reality and could foster misanthropy as an antisocial form of perverted "ecologism." Global bioethics, because of its bridging with economic and social reality, will be shown in this book to offer a more practical and yet no less spiritual alternative to "deep ecology," and because it considers human interests and responsibilities, it cannot be dismissed as being antisocial.

Anthropocentrism, Feminism, and Animal Rights

Other ethicists have addressed the question of our duties and responsibilities toward the environment from an "eco-feminist" perspective, notably Rosemary Radford Ruether in *Gaia and God: An Ecofeminist Theology of Earth Healing*[30] in which she calls for "biophilic mutuality," and Carolyn Merchant in *The Death of Nature*.[31] Carol Adams[32] was one of the first to apply a feminist perspective to the use and abuse of animals in contemporary society. J. Donovan and C.J. Adams have collected various feminist writers to explore the limitations of animal rights philosophy and an alternative ethics of care.[33] Considering certain parallels of prejudice in racism and sexism, Richard Ryder coined the term "speciesism" in reference to animals not being accorded rights.[34]

Utilitarian philosopher Peter Singer uses the criterion of sentience as a basis of moral worth. Singer's approach is essentially to bring animals into an extended, "preference" utilitarianism.[35] But the hierarchy of sentience upon which preferences are decided can be criticized for being anthropocentric, if not anthropomorphic. Tom Regan's orientation is nonutilitarian, making the case for animal rights on the basis of animals' interests and intrinsic value.[36] The contributions of these and other animal rights and protection oriented philosophers to advancing the moral status of animals cannot be adequately reviewed in the space of this book, notably the fine works of Bernard Rollin, Stephen Clark, Andrew Linzey, and Mary

Midgley that focus a virtue-based morality on our treatment of non-human animals.[37]

Notwithstanding Henry Salt's book on animal rights published in 1892, it was Albert Schweitzer with his ethic of reverence for life, along with increasing social concern over the welfare of animals and the plight of endangered species, that have drawn and inspired many scholars to address the question of animals' rights and human obligations.[38] Subsequent texts on animal rights were greatly enhanced by a deeper understanding of animal behavior and ecology that helped break down anthropocentrism, notably in the writings of David DeGrazia, Marc Bekoff and Dale Jamieson, Stephen Bostock, L. Finsen and S. Finsen, and R. Fouts and S. Mills.[39]

Midgley takes a rather different approach from other moral philosophers by considering how natural it is for humans to first care for those who are closest to them. She emphasizes that selfishness and anthropocentrism are not inherently wrong provided that in our concern for our own kind we recognize that "no man is an island." "From a practical angle," she writes, "this recognition does not harm green causes, because the measures needed today to save the human race are, by and large, the same measures that are needed to save the rest of the biosphere. There is simply no lifeboat utopia by which human beings can save themselves alone, either as a whole, or in particular areas."[40]

Bryan Norton observes:

> Long-sighted anthropocentrists and ecocentrists tend to adopt more and more similar policies as scientific evidence is gathered, because both value systems—and several others as well—point toward the common-denominator objective of protecting ecological contexts.[41]

Stephen Clark helps dispel anthropocentrism as follows:

> If we are to cope with our crisis, we must recognise the World as other than the human world, and recognise ourselves as inextricably dependent on that World. It is both our Other and our Origin, something unconstrained by our projected values and recognised as something by which we should be constrained . . . Moralists have tended to suggest that it is insofar as things are like 'us' that they are deserving of respect: but the better way is to respect them as not being ourselves, and so allow them to *be* . . . Of course, since

we are utterly dependent on the world (and so a minor part of it) there can be no gap between Us and It. But what It is does not depend on what we say it is. All attempts to evade this fact, like similar attempts to evade the laws of logic, seem to me to be appallingly misguided.[42]

The Empathetic Fallacy

One of the limitations of the animal rights/liberation philosophy and movement has been a lack of ethological and ecological understanding, which can result in what I call the "empathetic fallacy." The empathetic fallacy entails confusing sympathetic identification with empathic understanding of animals' feelings, needs and interests. This is sometimes evident in the actions and rhetoric of some animal liberationists and humanitarians. Their concern for animals is not always purely empathic or oriented towards justice and compassion. It may instead be confounded by anthropomorphic sentimentalism and by projection and transference when animals become symbols of social injustice and oppression or identified with as fellow victims. The empathetic fallacy can actually contribute to animal suffering, as when animal liberationists released mink from a fur farm in 1990 into the English countryside, putting both these animals and indigenous wildlife who fell prey to this carnivore at risk. The empathetic fallacy is evident in the rationalizations of some zoo people who contend that life in captivity is better for wild animals because they are safe from predators, poachers, climatic extremes, drought and famine.

I have on more than one occasion confronted the empathetic fallacy in my own veterinary work, in one instance concerning a severely burned dog. My immediate empathic response was that she should be euthanized because her burns were so extensive. But she showed so little discomfort while being treated that to consider euthanizing her had more to do with alleviating my own suffering for her than liberating her from her own suffering.

Those who sometimes experience the suffering of others so deeply that they actually manifest similar physical symptoms are certainly not caught in an empathetic fallacy. Nor are they necessarily suffering simply from a conversion hysteria. But the empathetic fallacy is evident when our identification with, and transference of, our feelings to others whose plight consumes us is so overwhelming that we become irrational, self-righteous, judgmental, and even violent. Examples include the fanaticism of religious and moral fundamentalists who have killed doctors, blown up their abortion clinics, and persecuted other doctors who engaged in euthanasia (mercy-killing).

Bringing Animal Rights and Environmental Ethics Together

J. Baird Callicott, in an earlier essay entitled "Animal Liberation: A Triangular Affair,"[43] drove a deep wedge between environmentalism and animal rights. He subsequently focused more on finding areas of common ground between environmentalism and animal rights where differences in moral theory could be reconciled.[44] But he still contends that a pluralism embracing both environmental ethics and animal rights is inconsistent, since "animal rights would prohibit controlling the populations of sentient animals by means of hunting, while environmental ethics would permit it."[45]

Bryan Norton comes to a similar conclusion that animal rights and environmental ethics can never be reconciled.[46] He criticizes those academics who, as spokespersons for deep ecology, have been able to avoid adopting policies on difficult, real-world cases such as elk destroying their wolf-free ranges, feral goats destroying indigenous vegetation in fragile lands, or park facilities overwhelmed by human visitors.

At one of the first conferences on animal rights in 1983, I urged that animal rights and the humane ethic "that are concerned almost exclusively with suffering, must be enlarged to incorporate nonsentient creations (plants, rivers, etc.) into an all embracing biospiritual or ecologically humane ethic."[47] Though seemingly absurd, it is possible to argue logically that rocks are worthy of moral consideration and accordingly due certain rights—primarily on the basis of their instrumental ecological, and inherent biological value (such as in preventing soil erosion and providing the soil and our food with trace-mineral nutrients). Bioscience can, therefore, provide an ethical basis for some rights arguments. These arguments make a synthesis of biology and ethics, i.e., bioethics. Similar bioethics-based rights arguments can be made for plants, animals, and microorganisms, and for the ecological communities they help sustain.

Bioethics, Animal Welfare and Animal Protection

In evaluating various uses of animals, and human-animal relationships from a bioethical perspective, we ask the following questions:

- In what way does a particular use or relationship benefit the animal?

- If the animal is harmed, then what human benefit can justify such harm?

- Are there alternatives that can give similar or greater human benefit without causing harm to the animal?

- If the kind of animal treatment or relationship does not satisfy the animal's basic physiological, behavioral and emotional needs, then what is the bioethical justification for so doing?

- If the kind of animal treatment, use, or relationship results in stress and distress, short-term or long-term, are the benefits that ultimately befall the animal greater or lesser than the human benefits arising from such use, treatment, or relationship? If they are less, then what is the ethical justification for such mistreatment?

Bioethics also calls us to ask how closely the nature of our relationship with the animal and the environment we provide for her accords with her *ethos* or intrinsic needs and nature; with her *telos* or biological purpose and fulfillment; and with her *ecos* or external environmental requirements (nutritional, spatial, social, security and comfort provisions) that are necessary for optimal physical health and psychological well-being. If there is any lack of accord that places the animal's health and well-being in jeopardy, then what is the ethical basis for justifying such a relationship, treatment or use? Is it actually to save a human life? Or is there a point of moral inversion or ethical inconsistency on the slippery-slope of animal exploitation where the costs to the animal, in terms of suffering, and to species and communities, in terms of extinction, cannot be justified for the benefit of individuals or society?

Where shall we make this point and draw the line between what is bioethically acceptable and morally reprehensible? The answer lies in the heart of reason that recognizes compassion and respect for all life as the basic bioethical principles of a just and humane society. These principles define what it means to be human, and when violated and animals suffer, we demean both our humanity and the sanctity of all life.

As will be shown in subsequent chapters, the interdisciplinary bridging of global bioethics makes a synthesis of environmental ethics and animal rights possible. Synthesis is important because there is no simple calculus or unanimity of human concern for nature and animals. There are many considerations, notably levels of: sentience (the capacity to suffer), intrinsic value (degree of selfhood or self-awareness), rarity (as of an endangered species), harmfulness and usefulness to humans, our own personal attachments (to our own family and companion animals), aesthetic value and sentimental attachment (liking elephants and whales), and cultural and religious significance (some rocks and cows are held sacred by some peoples, while for others, particular rocks and cows are of great monetary or totemic value).

It may be argued that all ethics by their very origin must be biased by anthropocentrism. Bioethics overcomes some such subjective limitations by appropriate reference to relevant objective knowledge from the sciences of ethology and ecology. A more biocentric or ecocentric orientation is thus achieved.

Global bioethics can also provide an antidote to what some critics, especially from the third world, call cultural imperialism and ethical colonialism, because the rights and traditional values of indigenous peoples are fundamental considerations. Similarly, concern for the rights, inherent value ("beingness") and biological and ecological values of animals as individuals, species and mixed communities are also fundamental considerations of global bioethics. This wider scope of ethical consideration helps overcome the animal rights/liberationist charge of "eco-fascism" against environmental and deep ecology ethics that put concern for ecosystems before concern for animals as individuals.

Conclusions

I have hardly done justice to the extensive and interesting literature on environmental ethics and animal rights, but I hope I have provided a useful overview for those who may wish to explore more academic treatises on these intertwined topics. So long as good ideas are put into action, the examination of our attitudes, values and relationships with animals and Nature will not become a self-cloistering substitute for bringing ethics to life. Moved by compassion, guided by reason, and inspired by passion, we can all find ways to serve the greater good, no matter what our station in life may be. The key for many of us, including myself, as we begin to incorporate bioethics into our lives, is to reflect upon the meaning of renunciation and communion and discover, in the heart's deep core of loving concern, the wellspring of enthusiasm (*entheos*, the God within) that will sustain us whenever our sense of commitment, duty or hope is in doubt.

My earlier attempts to integrate a diverse spectrum of animal rights and environmental concerns[48] led me to a "biospiritual" synthesis. Over the subsequent years of working on a variety of animal welfare and environmental (especially agricultural) issues, this synthesis had to be deconstructed. This was in part because of my discomfort with extremist animal liberation front activities; in part because of the recalcitrance of the corporate world to "go green" more deeply than as a public relations veneer; and because of the "new age" spiritualism that tended to trivialize deep ecology and animal rights with self-involved pursuits such as learning how to "channel." Some adepts claim to be able to talk to animals and trees, but do not take the next step to speak for them. The outcome of my own deconstructive process has been to construct a new, more holistic ethical paradigm (considerably inspired by Thomas Berry's *The Dream of the Earth*),[49] based on the fundamentals of empathy, reason, and biological realism. This new synthesis is what I see as the essence of bioethics: bringing life to ethics.

For some individuals, and humanitarian, animal welfare, and conservation organizations, bioethics may actually cause questioning of the kinds of projects, policies, and fund raising schemes that have made some non-profit organizations extremely well off and more focused on self-promotion and marketing than on ethically consistent and effective action. For the majority of people who do care about animals, nature, and human rights, bioethics will, I believe, enable a more effective engagement in a host of ethical issues than has been achieved by the animal liberation and Earth First! movements, and by the human rights movement. The first two have been hampered principally because they are seen by most people as extremist and misanthropic. The third, especially in its work concerning the rights of minorities and indigenous peoples, lacks impact because the lives of these groups seem so remote and unrelated to how most of us choose to live today.

Global bioethics, as defined and developed in this book, helps reconcile the polemics of animal (individual) rights and environmental ethics. This is in part achieved by cutting through the layers and hierarchies of human interests and values by providing biological realism (e.g., ecological and ethological knowledge) that has been deficient in earlier animal and human rights arguments and in environmental regulation and government oversight of industries and commerce. As will be shown in subsequent chapters, global bioethics also provides an empathetic and spiritual (Earth- or Creation-centered) framework for environmental ethics, biodiversity protection, and indigenous peoples' rights.

I have incorporated into bioethics the virtue ethics of moral philosophy. These differ from the two leading moral theories (utilitarianism or consequentialism) and rights theories (non-consequentialism, including duties) of today because they serve as a reflective template to evaluate and direct our aspirations, our behavior and the ways in which we treat other beings, human and nonhuman.[50] Virtue ethics are integral to the kind of global bioethics that we need to develop for our own good because they are reflective of how we choose to relate to each other, to other life forms, and to the living world. Virtue ethics can also be directive and achieve the same ends as utilitarian and rights arguments, but without the same intellectual baggage because virtue ethics are based on how we feel about ourselves and what it means to be human. I see virtue ethics as the human-centered component of global bioethics that, as philosopher David Hume contended, are grounded in altruistic feelings.[51] Charles Darwin proposed that such moral sensibility is a product of natural selection for the good of the community.[52] Stephen Clark, in a recent collection of essays, endorses the Aristotelean belief in the natural origin of moral law.[53] He points to our natural sentiments of sympathy and loving attention as being the natural roots of morality.

While bioethics is no panacea for the myriad problems, planetary and personal, that we face today, it does offer a way beyond the perceived polemics of human and animal rights and "deep" ecology, a way for us all to come together from very different points of view and vested interests. Under the same umbrella—or rainbow of global bioethics—as subsequent chapters will demonstrate, we can work together to make the world a better place for all life, and find ways to do more good than harm. But first we must deal with the question of putting people first before other animals and the environment—an issue that the next chapter addresses.

CHAPTER 4

People First?
Can Human Interests,
Animal Rights, and
Environmental Concerns
Be Reconciled?

Concern for human rights, for the interests and well-being of animals wild and tame and for the environment, leads us inevitably to conflict, as well as needless suffering and often pointless destruction. The rights of indigenous peoples to live traditionally, as hunter-gatherers, nomadic pastoralists or sedentary agrarians, like the native farming communities in Africa and South America and the family farms and ranches of Europe and North America, conflict with other economic interests. For example, the interests of industrial agriculture and its consequences, such as monopolistic control, and farm policies that favor mega-farms, are antithetical to the rights of many indigenous peoples the world over.

The natural rights of wild animals (to habitat, water access, grazing, etc.) have been long respected by most indigenous communities. An Indian peasant farmer, when I asked if he feared losing some of his stock to tigers that had been recently reintroduced in the jungle around his farm simply smiled and said, "The tiger is the spirit of the jungle, without which the jungle will never be well." Good farmers and ranchers, like native hunters and pastoralists, embrace this philosophy, not out of fear, but out of respect and understanding. This philosophy, based in part upon an attitude of assent—of live and let live—is the bedrock of all sustainable societies. The Indian farmer essentially pays rent to the tiger for using a few cleared acres of forest, knowing that the tiger will take some of his animals. That he has no desire to see the tiger eliminated is a lesson to us all. Yet the public pays for state and federal agents to kill predators on and around many farms and ranches in America. And in Tanzania, bushmen are fined and imprisoned for hunting in wildlife preserves—once the domain of such peoples for thousands of years. Likewise, the cattle of Maasai herders are often impounded if, out of hunger and thirst, they wander into a wildlife park or hunting preserve. They are not released until their owners have paid to enter the park and reimbursed the government for the costs of maintaining the cattle pound. While genocide continues—yes indeed, many Bushmen and Maasai communities who still cling religiously to their traditional way of life are so impoverished that they are on the verge of starvation and extinction—ecocide intensifies. The land is being divided up, and wildlife communities and species are being wiped off the earth.

Humane and conservation organizations that would divide up the land to exclude such peoples often do more harm than good because some studies have shown that these peoples have helped shape the land, and even increased natural biodiversity. The Bushman lived in harmony with wildlife for some 100,000 years and more. And for centuries the Maasai never killed wild animals for their meat, choosing to live in balance, along with their cattle, sheep and goats on the plains of Africa. That they are now being excluded in the name of wildlife protection and conservation from the very land that they have long stewarded, as conservators and protectors, points to a significant lack of vision and understanding by those with the power to change their lives.

Some try to overcome this conflict by setting up "People First," or "Earth" or "Animal First" organizations. If these help at all, they show us how ethically inconsistent and confused we are, and as some would say, how morally bankrupt and spiritually blind we have become. Certainly I would celebrate the demise of all the ranchers who oppose the reintroduction of the wolf into Yellowstone National Park. These ranchers should go live with the Indian farmer who respects his jungle of tigers, and learn something. In the process, they might come to earn the support of those conservation and animal protection organizations that care as much for the family farm and ranch as they do for wolves, Indians and buffaloes.

Farmers and ranchers, among others, have been indoctrinated with the notion that the "tree-hugging, Bambi lovers" are out to put them out of business. What's really putting them out of business are those who fund such polemicizing organizations as the Animal

Industry Foundation, Putting People First, the Wise Use environmental movement, and the Americans for Medical Progress Education Foundation. Be they ideologically pure or conspiratorial, these organizations are creating a public smokescreen that has led many farmers and ranchers to blame some animal protection and conservation organizations for their economic demise. Yet, these and other good societies are confronting the very same forces that threaten the way of life of indigenous peoples worldwide.

That some Maasai have killed tourists in Kenya in protest of measures by the Ministry of Tourism to restrict their access to traditional grazing lands, points to how bad the conflicts are becoming today. That three or four companies now have a virtual monopoly over the U.S. livestock industry, in spite of anti-trust laws designed to prevent such a centralization of power and control, is cause for concern. So is the loss of half a million family farms in the U.S. over the past decade. Corporate imperialists are now manipulating the General Agreement on Tariff and Trade (GATT) convention as they have the North American Free Trade Agreement (NAFTA) to favor their own ends. This may well result in a new world order of corporate feudalism if it is not conceptually realigned with the principles and morality of a sustainable and just society, economy and agriculture.

Bioethics is the key to finding alternatives and solutions. The trans-situational attitude of respect and reverence for all life comes from embracing the ethic of *equal concern* (equalitarianism) for animals, environment and people alike. No one is first. What must come first from us is compassion. This is the essential ingredient of a viable life ethic.

The boundless ethic of compassion expressed in reverential respect for all life can serve as a foundation on which to build a bioethical framework for humankind—a framework that helps preserve community integrity and cultural diversity through defense of human rights, as much as it protects biological diversity by upholding the humanitarians' and conservationists' claim that all living beings have a right to live in a whole and healthy environment. With such an ethical foundation there will then be no valid reason for anyone to claim that people or animals come first, or desire to put jobs before environmental protection and conservation, because all acknowledge that what harms the environment inevitably harms people. And when we come to condone cruelty toward animals, we also tend to become indifferent toward the suffering and injustices of our own kind.

Bioethics as a Philosophical Integration of Human, Animal, and Environmental Rights

It is considered heroic altruism for men to help women and children first onto lifeboats of a sinking ship. But what would Noah have done if he, his family, and the world's menagerie of creatures were on a sinking ark? Ecologist Norman Myers in his book, *The Sinking Ark*, shows very clearly that both we and our animal kin are in grave danger of extinction, and we have no lifeboats or safe place to go.[1] Furthermore, our heroic altruism to save our own kin and species, as ecologist Garret Hardin explains in his book *The Limits of Altruism*, can be misguided.[2] We ultimately harm ourselves in the process of valuing human life over the "rights" of other living beings and in ignorance of sound ecological principles. These principles include what Hardin terms the "carrying-capacity ethic" of not living or multiplying beyond nature's capacity to sustain our needs and the needs of other living beings that play an essential role in the maintenance of a healthy environment.

Humanitarians and conservationists, when they voice concern over the fate of animals, wild and tame, are mistakenly seen by their critics as placing animals before people. In the critics' minds, such concern is variously interpreted as anthropomorphic or sentimentally misguided altruism, since no rational person would value animal life over human life. Biomedical scientists and spokespersons for various industries, such as agribusiness, energy, and forestry, are especially critical of humanitarians and conservationists. To question the exploitation of animals and ecosystems is to be heretical, anti-progress, a neo-Luddite. But there are surely reasonable limits to such exploitation, and those limits have yet to be defined and agreed upon both nationally and internationally.

A dialectical tension is evident between our need to exploit and destroy life in order to sustain our own, and the desire to respect, cherish, and preserve all life from wanton exploitation and suffering. This results in a polarization which some moral philosophers have attempted to resolve by arguing that animals have intrinsic value and that there are as yet no empirically demonstrable, morally relevant differences between humans and other animals to justify their wholesale exploitation for often trivial purposes. Might (dominion expressed as domination) does not make right.

No matter how articulate and airtight these moral arguments, the fact remains that most people are convinced that animals are inferior, if not created for our own exclusive use. To suggest that

animals should be given equal and fair consideration is interpreted as placing animals before people. Given our anthropocentric cultural values and our intellectual and technological
superiority over other animals, it is easy to believe that we are in
certain ways superior. This may be true in some aspects, yet in
terms of being able to survive by wits alone in the desert, the coyote is by far our superior.

The benefits—social, economic, and moral—of animal rights
philosophy and ecological ethics will never be realized until the
polemic of "animals before humans" versus "humans before animals" is reconciled. Intellectual argument, appeal to reason or to
higher moral and spiritual values, can be too easily dismissed on
the grounds of utility and necessity. Once there is economic or
other dependence, all forms of animal exploitation and environmental manipulation and destruction are justified and become normative. Moral argument is then relatively futile, especially in
cultures that place economics (jobs and profits) before ethics and
thus put the interests and rights of people before animals and the
environment.

A major limitation of animal rights philosophy is that it tends
to foster self-righteous moralizing, a sense of moral superiority, and
a judgmental attitude toward opponents which only increases the
existing polarization. Furthermore, intellectual argument simply
meets its own resistance in the form of counter arguments and rationalizations. No matter how clearly presented, animal rights philosophy and environmental ethics will continue to be resisted by
those who accept that animal and environmental exploitation is
necessary for the human good, and by those who believe that those
who speak in defense of animals have neither concern nor compassion for people. Certainly there are extremist factions who put
animals before people or people before animals, and their con-
tentiousness is escalating today. Such polemicizing fundamentalism
aside, the more moderate and reasonable view is to reconcile this dialectical tension by acknowledging that both factions are probably
half right.

So where do we go from here? A philosophic integration of human, animal and environmental rights is needed, and bioethics
provides the core concept for this necessary integration. Such an
integration will provide a more democratic and compassionate basis for determining public policy and personal behavior in terms of
our regard for and treatment of animals, the environment, and
each other.

Reconciling Differences

People who do not appear to have much, if any, concern for animal rights and environmental issues are so inclined for many reasons: ignorance, giving priority to immediate survival needs, custom, religious and secular value systems, lifestyles, economics, and especially a lack of feeling. This is not to say that people lack a capacity to care, only that that capacity has not yet been informed and awakened. Feeling empathy for animals and nature is a function of parental example, older peer pressure, school, and religious education.

People who have no feeling for animals and the environment (other than as personal subjects of sentiment or objects of utility), cannot be expected to care about animal rights and environmental ethics. They may even actively seek to discredit the legitimacy of animal protection, conservation, and environmental quality issues. This is especially true when the proposed correctives threaten to undermine the status quo of industry, religious and secular traditions, political ideology, or the economy. The establishment's fears and opposition to the linked concerns of environmental ethics, human rights and animal rights are cause for reflection.

Ethical Colonialism and Cultural Imperialism

The imposition of one's own values on others can amount to ethical colonialism and cultural imperialism and is often judged as racist. The basic bioethical principle of equalitarianism can be difficult to apply in practice, as witness the 1999 controversy over the legally permitted gray whale hunt by the Makah Tribe of Washington State. Not all tribal members were in support of this hunt under rights granted to the Tribe in an 1855 treaty.[3] But supporters claimed that hunting and killing whales was their right, part of their cultural heritage and would help give their youth a sense of pride and cultural identity. They called anti-whale hunt animal rightists and environmentalists racist, and failed to acknowledge the legitimacy of public concern for the welfare of whales, which are declining in population. The pro-hunting tribal faction also refused to acknowledge that other Indian tribes like the Quilliute and Yaqui are quite able to continue their cultural rituals and maintain their cultural identity without killing totemic animals as they did in the past. This is a case in point of anthropocentrism and selfish disregard for a sacred, totemic animal, the whale. Had the pro-hunting

Makahs considered the needs of the whale in and for itself, they would have opposed the hunt and joined animal rightists and environmentalists in protecting whales' rights, not killing them in the name of cultural identity.

Cultures must either evolve or become extinct. Evolution in this instance would have meant identifying with the endangered status of the gray whale and finding cultural identity, pride, empowerment, and renewal in being its protector. In killing the whales—in nontraditional, if not sacrilegious ways according to eye witnesses—the Makah whale killers opened the door for other whale killing nations like Japan to follow their claim of cultural heritage and indigenous peoples' rights to continue commercial whaling.

Another instance of potential ethical colonialism and cultural imperialism worthy of some reflection concerns the imposition of dietary habits of one culture upon another, or the ridicule of a culture's dietary habits. For example, we have the Hindu's aversion to Muslims' and Christians' killing of cows and their calves for meat; the British abhorrence of the French killing and eating of horses; the western abhorrence of Vietnamese and Chinese and other occidental countries' raising and killing of dogs for their meat. Within the diverse Hindu culture of India, there is abhorrence and opposition to the continued ritual slaughter of animals in temples, paralleling the opposition in many western countries to the ritual mutilation and slaughter of bulls in the ring in countries like Spain, Portugal, and Mexico. Other disturbing practices include the genital mutilation of women, hand amputation of thieves, human bondage, and extreme forms of child labor and exploitation that are still socially acceptable in some countries and communities.

As we move toward increasing economic interdependence and interaction between different cultures, we encounter a variety of different ethical positions, born from different cultures' traditions, and values. Applying bioethics globally to facilitate mutual respect and understanding, as well as constructive nonjudgmental dialog, will contribute significantly to helping cultures learn from each other the virtues of compassion, humility, and reverence for life, and in the process evolve. Applying bioethics means that traditional abuses of sentient beings, including humans, for religious and other reasons will be increasingly questioned and outlawed. Fighting these practices will no longer be judged as ethical colonialism and cultural imperialism when those involved are in the spotlight of a more compassionate and caring, collective world conscience.

When people have no concern for the environment and other sentient life, and refuse to hear the reasoned logic of bioethics, then we who are concerned must address their fears, hopes, prejudices, values, and motivations, and work together as best we can to keep justice and compassion alive in our global community. The essential virtue of bioethics is that it links altruism with enlightened self-interest. Fascism, racism, sexism, and "speciesism" aside, whatever utopian vision or hopes for the future we may have, we should all consider how our actions, values, policies and goals might harm or benefit. We can begin by completing a personal inventory. Governments and corporations ought to conduct bioethically-based impact assessments of all past, present, and planned actions, policies, products, and services. This is not an unreasonable proposal, nor is it idealistically utopian and impractical. It is surely relevant, reasonable, and the essence of humane responsibility. A more penetrating evaluation of the role of bioethics in conserving wildlife and biodiversity, and transforming human-centered values and belief systems in different cultures is now appropriate and is the essence of the next chapter.

Bioethics of Conservation

That which separates us from other animals is our biblical power of dominion. Our dexterity and instrumental knowledge enable us to transform matter, life forms, and whole ecosystems. Through millions of years of co-evolution, nonhuman animals have been constrained by the laws of Nature, the regulatory processes of the ecosystem in which they maintain their niche and fulfill their role or purpose. By contrast, the human species has, in the process of expanding its niche, assimilated and annihilated other species and communities, both human and nonhuman. One may wonder what purpose or role this serves in Earth's Creation.

Some traditional peoples, like native Americans, hold that we should imitate animals, who are obedient to their instincts, and follow our "original instructions" which, according to native American spiritual cosmology, mandate living in balance and harmony with other beings and nonhuman communities. This ethic is anathema to conventional wisdom. Regardless, unlike other animals whose power of dominion (or domination and control) is far less than ours, we can choose how best we should use this power. We can choose to live gently, in respect and reverence for all life, and in obedient understanding of the laws of Nature, or we can choose to live otherwise, bringing imbalance and disharmony into the world. This is not to imply that some animal species are lacking in any moral sense or empathic sensitivity. Rather, we must recognize that while we have the freedom and power to act as though we are no longer subject to the laws of Nature, it is enlightened self-interest to be obedient to these laws because we are ultimately subject to them. This power and freedom brings with it the enormous responsibility as moral beings to temper our instincts and desires. Choice mandates moral discrimination, and it is from the synthesis of our emergent moral sensibility with our understanding of the laws of Nature that the life ethic or bioethic of a humane, socially just and sustainable society is derived.

A purely utilitarian, exploitative attitude toward animals and the natural world is morally reprehensible in an affluent society, but is understandable in bioregions where there is poverty and near famine. In the face of abject poverty, aggravated by escalating over-population and dwindling natural resources, the human survival instincts for procreation and satisfaction of basic food, water, and shelter needs take precedence over those of wildlife and biodiversity, conservation, and protection. Given this, it is bioethically imperative to prevent humanity from spiraling downward into increasing poverty and suffering, environmental degradation, and species extinction—a condition in which we find many African countries. The possibility for an empathic concern and reverence for animals and Nature becomes ever more remote when our basic needs cannot be met without destroying other species—even ourselves—and the life-support systems of the bioregion. In other words, when their own lives are imbued with the hopelessness and fatalism of abject poverty, most people cannot spare concern for others, and ethical sensibility remains latent.

Every child should be taught more than a purely utilitarian or superficially aesthetic attitude toward other living beings, including wildlife. When children are not taught to appreciate the sanctity of all life and our sacred duty to protect, as much as we must exploit to satisfy our basic needs, the socialization process is impaired. This impaired process is well described in A. Richard Mordi's *Attitudes Toward Wildlife in Botswana*.[1] The end result is overpopulation, human suffering, and the annihilation of wildlife and the natural world.

Even with massive international aid and development programs, in some regions it seems too late to reverse the downward spiral of poverty, overpopulation, and environmental degradation, and to prevent the extinction of wildlife such as the elephant, hippo, rhino, wild dog, and cheetah. These programs will continue to fail and cause more harm than good for as Mordi's study indicates, they are not coupled with a concerted effort to change people's attitudes toward wildlife and nature, or provide them with the technology and information that would enable them to adopt conservationist agriculture practices and attain self-sufficiency and zero population growth.

The notion that economic development and industrial growth via resource commoditization is the best strategy for global conservation and for overcoming poverty and hunger is now being embraced by several international conservation organizations. Other conservation organizations refuse to embrace this notion of economism, or economic determinism, and are suspicious of alliances between transnational banks, multinational corporations, and international conservation organizations, such as the World Wildlife Fund for Nature. Aside from the ethics of applying monetary value to wildlife, genetic plant resources, and the medicines of indigenous peoples, the process of establishing markets for such products, including eco-tourism and safari hunting, could harm the interests and goals of conservation.

First, putting a purely monetary value on elephant ivory and other wildlife products will encourage poaching. It will also undermine attempts to establish questionable sustainable harvesting (where killing-quotas are set and policed.) Second, the benefits local peoples might receive from marketing wildlife, tourism, and hunting could quickly evaporate, given the volatile nature of world markets and the global economy. Local communities would then be left with bankrupt game parks, which would quickly be decimated if these people have not been helped to develop their own viable social economy, and especially sustainable agricultural and resource management practices.[2] Aside from the issue of indigenous peoples' intellectual property rights and interests, economism could severely undermine more sustainable cultural traditions. For example, encouraging them to rely on food imports and to use their land to raise cash crops for export would put the entire community at risk from external market forces over which they have no direct control. If these concerns of equity and cultural integrity are not given due consideration, a purely commercial approach to wildlife conservation and protection of biodiversity is likely to cause more harm than good.[3]

Once people are set in their ways and driven by poverty, changing attitudes is extremely difficult, if not impossible. Hence, the importance of early education to socialize children with wildlife and ecology through direct encounter if possible, so enabling them to understand and appreciate the intrinsic value and ecological role of animals and plants in maintaining the character, quality, functional integrity, and bioeconomy of their region. A lack of such understanding and appreciation is invariably coupled with the erroneous belief that wild animals can take care of themselves and that Nature is self-regenerating and can always withstand and accommodate human impacts. Environmental disasters, so often anthropogenic, like droughts and floods following deforestation and overgrazing, are likewise erroneously seen fatalistically as being "natural"—perturbations of an untrustworthy Nature. Or else they are seen as supernatural retribution, rather than caused by the people themselves. But when the socialization process is not impaired, a more ecocentric or biocentric perception and understanding is possible. Solutions can be developed and further ecological degradation, species extinction, and human suffering can be prevented.

That one man's gain can mean another's pain is the zero sum of progress in the postmodern age. Overpopulation, overconsumption of finite resources, and environmental, moral, and spiritual deterioration are realities of the times that cannot be denied. That we are increasingly violent to each other, to animals, and to Nature, points to a dysfunctional human condition that needs to be acknowledged and its primary cause understood. Mutual distrust (xenophobia), cruel indifference toward lesser beings, and our very conception of self must, for the health of the populace and planet alike, be superseded by mutual aid (altruism), compassionate reverence for all life, and full self-realization (see Table 5.1).

Realizing the true nature of the self means more than understanding the nature of reality and overcoming cultural distortions in how we think, feel, and sense. It is neither an esoteric or a mystical process. It is simply the process of determining how well socialized we are to each other, to Nature, and to all sentient beings. The more socialized we are—spiritually, ethically, and empathically connected with each other and the Earth—the more civilized and self-aware we become. It is especially during childhood and adolescence that this

Table 5.1 The Self-Realization and Socialization Process

Developmental Direction Influenced by Parents, Teachers, Resource Availability, and Other Factors	SELF ⟶ OTHER	
	Self-preservation	Resource conservation
	Self-reliance	Community interdependence
	Self-respect	Affirmation from others
	Self-control	Respect for others
	Self-love	Compassion toward others
	Self-confidence	Mindfulness of others
	Self-determination	Freedom of expression
	Self-sacrifice	Mutual cooperation
	Self-awareness	Social conscience
	Self-realization	Community actualization

socialization process takes place. When it is impaired, the individual and the community suffer the consequences. When the community-at-large is comprised of individuals who have not been socialized, their offspring and the life community of the world they inhabit suffer the consequences.

Poor third-world farmers' and graziers' antipathy toward wildlife that consume their crops and livestock is understandable. Yet the rage and prejudice toward predators, particularly among some ranchers in the U.S. and other developed countries, is disturbing because one would expect a greater degree of education about wildlife and ecology among sheep and cattle ranchers in these countries than among those in the third world. But this is not the case. Among some ranching communities there is still a palpable fear and loathing of wolves which reveals an attitude that has played a major role in the degradation of rangeland and consequent economic problems that these people have brought upon themselves to a large extent. Emotions reflect beliefs and values, and these negative emotions toward wildlife, particularly predators, reflect a degree of ignorance that is incompatible with the principles and practice of humane, sustainable ranching and agriculture. That the public has paid for government agents to exterminate wildlife for many decades on public lands, and continues to subsidize farming, ranching, mining, and timber industries that are destroying the last of the wild, is morally reprehensible and economically disastrous.

In order to avoid judgmentalism, it is ethically wise to separate the act from the actor. Thus we might feel that seal killers are not evil, but clubbing seals to death for their pelts is evil. Making these pelts into coats and wearing these coats are, by extension, also evil. But in separating the act from the actor, we discount the element of choice. People have freedom of choice, no matter what taste, fashion, tradition, and financial need dictate. Furthermore, no matter how diplomatically we might voice concern over inhumane acts, the actors feel blame, if not also guilt, because they are the perpetrators of what others judge as evil. Causing harm is often deemed acceptable and necessary for achieving certain ends, such as for nourishment and control of pests and disease. But as our ethical sensibility evolves (i.e., as we become more human/humane), we strive to find less harmful alternatives and forsake such trivial, nonessential ends as fur coats, ivory ornaments, and goose paté where the means entail harming other sentient beings or the environment.

There are no simple solutions to the moral complexity and ethical dilemmas that we face in addressing many contemporary issues.

A broader, more holistic bioethical approach is needed, which Albert Schweitzer envisioned as follows:

> A man is ethical only when life, as such, is sacred to him, that of plants and animals as that of his fellow men, and when he devotes himself helpfully to all life that is in need of help. Only the universal ethic of the feeling of responsibility in an ever-widening sphere for all that lives—only that ethic can be founded in thought. The ethic of the relation of man to man is not something apart by itself: It is only a particular relation which results from the universal one.[4]

Leopold's vision of the gradual evolution of a broadening ethical sensibility has been schematicized in two drawings by Roderick Frazer Nash in his excellent book *The Rights of Nature: A History of Environmental Ethics* (See Figs. 5.1 and 5.2)

Nash explains these figures, and presents documented historical evidence as to their fidelity, as follows:

> Figure [5.]1 attempts to show what exponents of evolved or sequential ethics believe. The time line along the figure's left margin suggests that ethics awaited the development of an intelligence capable of conceptualizing right and wrong. And even then, for long periods of time, morality was usually mired in self-interest, as for some it still is. . . .
>
> The second drawing, Figure [5.]2, is a schematic view of the historical tradition of extending rights to oppressed minorities in Britain and then in the United States. At the center are the natural rights tradition and the concept of intrinsic value that date to Greek and Roman jurisprudence. The diagram lists the key document that codified each new minority's inclusion within the circle of ethical consideration. Figure 2 does not imply that the minority immediately attained full rights in practice as well as theory on the given date, nor that only the documents listed were important in establishing minority rights. Its purpose is merely to show that ethics have expanded over time and that some thinkers and activists now regard nature (or certain of its components) as deserving liberation from human domination. For people of this persuasion natural rights has indeed evolved into the rights of nature . . .[5]

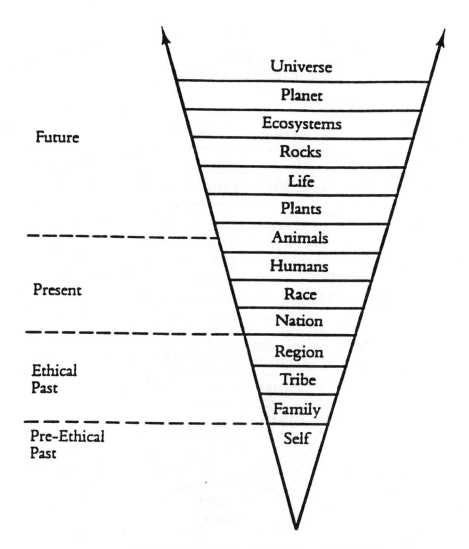

Fig. 5.1. The Evolution of Ethics (from Nash, 1989).

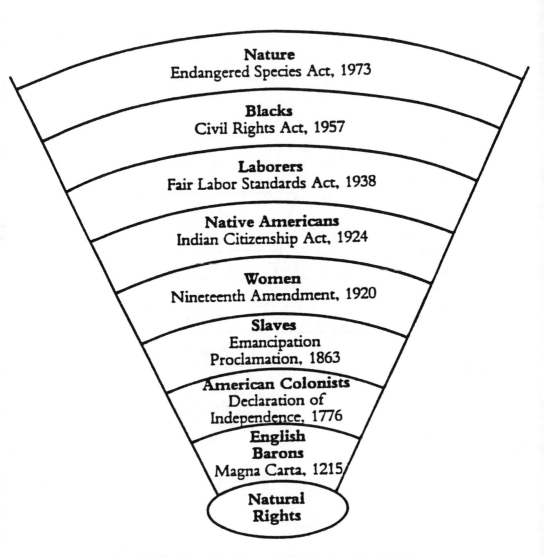

Fig. 5.2. The Expanding Concept of Rights (from Nash, 1989).

From Nash's detailed analysis, we can conclude that the seeds of hope have indeed been sown, and even in the face of increasing global problems, continue to flourish. As the most consciously transformative species on Earth, a responsible choice on behalf of humane planetary care and service is surely the essence of human purpose and fulfillment.

E. O. Wilson has coined the term *biophilia* with a book of that title. He defines biophilia as "the innately emotional affiliation of human beings to other living organisms."[6] The conclusion I draw from Wilson's work is optimistic: to the degree that we come to understand other organisms, we will place a greater value on them and on ourselves. In this sense, biophilia is an antidote to the accelerating loss of biodiversity and extinction of plant and animal species and communities.

In many cultures, however, the barriers to biophilia are considerable, if not insurmountable. In cultures where the prevailing attitude toward life is utilitarian, biophilia is as alien as the notion of reverence for all life. But since most human societies have some codes of ethics and morality, these can be built on and expanded to include consideration of plants, animals, and nature. Where the prevailing attitude is utilitarian, linking bioethics with enlightened self-interest will facilitate the transition from selfish exploitation to a responsible and compassionate stewardship.

I term biophilic knowledge of the biology of organisms and ecosystems *biosophy*, the reward of biophilia. But as Wilson cautions, it is difficult to decide what is best for both the near and distant futures. Because short- and long-term decisions are often internally contradictory, ethical codes need to be formulated. Biosophy takes us beyond Wilson's contention that the only way to make a conservation ethic work is to ground it in enlightened selfishness. Certainly people will conserve land and species if they see a material gain for themselves, but such enlightened utilitarianism is a beginning, but not an end in itself. It neither engenders nor fosters social justice, or compassion, or respect for the inherent worth and rights of living beings. Furthermore, in many regions where wildlife is critically endangered, as in Africa, a predominantly utilitarian attitude has played a major role in the demise of wildlife and habitat. In these decimated regions, the human spirit has been crushed by poverty, corruption and despair. Wildlife and resources are owned by governments and a wealthy elite that have little regard for the poor and the disenfranchised and see material gain for themselves, their kin, and their tribe in wildlife and resource exploitation.

The bioethical formulations that arise from biosophy can provide the epigenetic framework for human evolution to better ensure the preservation of the human spirit and the life and beauty of the planet. Biosophic tenets along with the Precautionary Principle[7] can help us get out of the evolutionary dead end of selfish exploitation so that we conserve Nature for Nature's sake, and other living beings for their own sake. But in these worsening times, environmentalist John Terbough is right I believe in calling for the deployment of well armed international "nature keeping forces" akin to United Nations peacekeeping forces, if we are to save the last of the wild in most countries.[8] Our moral and ethical maturity will be evident when we begin to act as though we belong to the world and not as though the world belongs to us. We must also accept that there is more than one way to live. Self-determination is a right that is conditional upon living as harmlessly as possible according to the principle of ahimsa, and with an assenting attitude of "live and let live" with regard to other cultures, religious traditions, and fellow creatures in the wild.

A Life Ethic that promotes respect for all life arises from an understanding of the wisdom of life, biosophy, as distinct from the objective study of life, biology. Ethology, the study of animal behavior, is an important area of biology, enabling us to better understand the ethos, the inner life and expressive nature, of animals. Ethology has considerable application in developing humane husbandry practices for domestic animals, and humane treatment for wildlife. As African ecologist and teacher Baba Dioum observed, "In the end we will conserve only what we love; we will love only what we understand; we will understand only what we are taught."

It is from the perspective of biosophy that the bioethical principles of the Life Ethic are identified and articulated. For those familiar with theosophy, as distinct from theology, biosophy is a complimentary field of human inquiry that grounds theosophy in the here and now. It moves us to live in accordance with a creed of compassionate concern and reverence for all life. It calls us to action whenever there is injustice and inhumanity and inspires us to avoid harming other sentient beings through acts of omission and commission. The biosophical revelation that it is enlightened self-interest to embrace ahimsa (non-injury) and active compassion is the wisdom of life that has been recognized from age to age. While biosophy needs no theology, it is not unscientific. It is surely a most objective science of responsible and appropriate human behavior cast within the realistic dimension of our place and responsibilities toward the life community of this planet.

Of all human activities that have historically caused great harm to the environment, to wild and domestic animals, and ultimately to the human community, agriculture ranks first and foremost. Agriculture, as the next chapter shows, continues to be a major threat to conservation and biodiversity, and is cause of much animal suffering and unnecessary consumer health risks.

CHAPTER 6

To Farm Without Harm and
Choosing a Humane Diet:
The Bioethics of
Sustainable Agriculture

No other society past or present raises and kills so many animals just for their meat. No other society past or present has adopted such intensive systems of animal production and nonrenewable resource dependent farming practices. These have evolved to make meat a dietary staple and to meet the public expectation and demand for a "cheap" and plentiful supply of meat. An agriculture that raises and slaughters billions of animals every year primarily for meat depends on costly nonrenewable natural resources and precious farmland to raise the feed for these animals to convert into flesh—land that critics now believe should instead be used more economically to feed people directly.[1] To a hungry world, such conspicuous consumption is a poor model to emulate.

Supporters of intensive animal factory farming claim that America has the cheapest and most productive agriculture in the world and that humane reforms would increase costs and put an unfair burden on the poor. Critics of factory farming are judged as being more concerned about animals than people and opposed to progress. Both these erroneous beliefs and conclusions need to be dispelled.

The real costs of factory farming have been well documented, ranging from price supports and subsidies at taxpayers' expense, to the demise of family farms, rural communities, waste of natural resources and animal stress, disease, and suffering.[2] Coupled with corporate monopoly, these hidden costs have aggravated rather than alleviated poverty and malnutrition nationally and internationally.

The real costs of factory farming are not accounted for by agribusiness, and its high productivity is neither efficient nor socially or ethically acceptable. Some of the reasons for reaching this regrettable conclusion will now be detailed.

Bioethical Travesties of the Livestock Industry

For those not familiar with the bioethical travesties of conventional livestock production, the following concerns will provide a brief introduction.

- Extreme confinement and animal suffering are the norm. Veal calves and sows are unable to walk or turn around. Laying hens are confined to "battery" cages, 4–5 birds living in a space too small for even one to stretch her wings. Dairy cows and beef cattle are confined to dirty feedlots with no access to pasture and often no shade or shelter. These conditions cause animals distress and stress and create an ideal environment for the spread of so-called production or husbandry-related diseases that cause animals further suffering.

- The antibiotics and other drugs that farmers make widespread use of to control stress-related diseases and to stimulate growth, and egg and milk production, leave residues that pose health risks to consumers, and cause the development of antibiotic resistant strains of bacteria responsible for food poisoning epidemics.

- Farms squander natural resources, notably arable land and water, to raise corn and soybeans for livestock feed that is inefficiently converted into animal fat and protein. Animal wastes (poultry manure fed to cattle) and the condemned and unused remains of slaughtered livestock (44 billion pounds per year in the U.S.) are included in livestock feed, as well as the rendered remains of euthanized cats and dogs and road kills. In Europe, the use of animal remains in animal feed led to bovine spongiform encephalopathy or "mad cow" disease.

- The billions of livestock in the U.S. and other industrialized countries produce vast quantities of urine and feces, too much for the surrounding farmland to recycle, the net result being surface and groundwater pollution, fish kills, bacterial, and drug contamination of drinking water, and contribution to global warming.

- Animal wastes, including diseased offal that when dumped at sea cause ecological damage, may cause diseases in marine wildlife and pelagic avifauna.

- Herbicides, pesticides, synthetic fertilizers, fungicides, and new, genetically engineered crops used by the livestock and feed industry cause environmental and genetic pollution, endangering wild plants and animals and contiguous organic farming systems.

- Predaticides, from poison baits to cyanide guns and trapping, coupled with overstocking and overgrazing by livestock, decimate natural ecosystems where cattle are bred and raised prior to going to feedlots to be "finished" for human consumption.

- The above activities are encouraged by the government and agribusiness system through loans, subsidies, and tax write-offs, all at taxpayers' expense and at a significant cost to public health. This benefits the biomedical animal research and pharmaceutical-medical-industrial complex, that now calls itself the life science industry.

- The "bioconcentration camps" of the animal industry have ruined the sustainable economy and livelihoods of family farms and rural communities through market monopoly. In developing countries they have had a similar impact, first by coopting land owners to produce livestock feed for export, and second by encouraging and subsidizing the adoption of intensive poultry, pig, beef, and dairy production systems. An additional problem is "dumping" of surplus livestock and other agricultural produce on developing countries. This puts local producers out of business because their production costs are higher than the market price of these "dumped" imports—like U.S. powdered milk and chicken parts in Jamaica. Indigenous peasant farmers in some developing countries have even been forced off their land at gun point by the military whose governments want their land to raise soybeans for export to the U.S. livestock industry.

- The net consequences of colonial agribusiness are manifold: indigenous farmers are bankrupt and disenfranchised; indigenous knowledge, seed-stocks, and sustainable agricultural practices—the keystones of biocultural diversity—are lost; rural communities are impoverished and malnourished, and have to rely on food imports, often of inferior quality; many emigrate to urban slums to seek employment, while others degrade marginal lands and encroach on wildlife preserves to graze livestock and raise food crops, and poach bushmeat and various forest products, practices often encouraged by corrupt authorities, who are supported by international donor agencies.

- People in developing countries who can afford to adopt the western diet high in animal fat and protein soon develop western diseases associated with such a diet, notably arteriosclerosis, osteoporosis, and various forms of cancer. In the process they unwittingly support one of the most harmful and costly industries in the world—the livestock industry which, along with agri-biotechnology and fast-food franchises like McDonald's and Kentucky Fried Chicken, are encouraged by the U.S. State Department through their embassies around the world. (See Table 6.1.)

The Harms of Overproduction

In some countries, like Brazil, raising livestock has become a major hedge against inflation, but overproduction cycles depress world market prices, and fuel deforestation and other forms of environmental degradation. Price supports and subsidies to producers, especially in the developed world, encourage overproduction and cause further distortions and inequities in world market prices. One serious consequence is the "dumping" of meat, dairy, and other agricultural products in other countries—products that are then sold to processors and wholesalers at prices much lower than local farmers can get for their own similar produce. Import tariffs to help protect local farmers from dumping and from being forced out of business further compound the problems of agricultural surpluses and subsidized export commodities coming from more industrialized nations. While tariffs and other forms of protectionism by any country to protect its own farmers is an illegal "technical barrier" under the GATT convention, local farmers raising food and feed for domestic con-

Table 6.1. Western Diseases of Known
and Possible Dietary Origin*

Gastrointestinal—Constipation, Hiatus hernia, Appendicitis, Diverticular disease, Colorectal polyps, Crohn's disease (regional ileitis), Celiac disease, Peptic ulcer, Hemorrhoids, Ulcerative colitis.

Cardiovascular—Coronary heart disease, Cerebrovascular disease (stroke), Essential hypertension, Deep vein thrombosis, Pulmonary embolism, Pelvic phleboliths, Varicose veins.

Metabolic—Obesity, Diabetes (type II or noninsulin dependent), Cholesterol gallstones, Renal stones, Osteoporosis, Gout.

Cancer—Colorectal, Breast, Prostate, Lung, Endometrium, Ovarian.

Autoimmune diseases—Diabetes (type I or insulin dependent), Autoimmune thyroiditis.

Other disorders—Allergies, Immunoinsufficiency, Infantile hyperactivity, Migraine, Multiple sclerosis, Pernicious anemia, Rheumatoid arthritis, Spina bifida, Thyrotoxicosis.

*Modified after D. P. Burkitt in *Western Diseases: Their Dietary Prevention and Reversibility*. N. J. Temple and D. P. Burkitt (eds.). Totowa, New Jersey: Humana Press, 1994.

sumption should have their market protected and fair market prices guaranteed, provided their farming methods are humane, socio-economically just and ecologically sound and sustainable. And they should not be encouraged to adopt the capital intensive, high-input methods of animal agriculture that have become the bane of the industrialized world.

The high-volume productivity of industrial-scale, intensive systems of livestock and poultry production is often touted as being the hallmark and miracle of progress and success. Poorer, developing countries are encouraged to adopt these methods in order to increase agricultural production and efficiency. Yet ironically, the global industrialization of animal agriculture is now counterproductive in part because it is too successful. Industrialized countries are passing on the burden of overproduction and commodity surpluses to the third world while at the same time their industrial agricultural experts, agribusiness agents, and development banks are trying to sell intensive livestock and poultry production systems to these countries.

The legal definition of "dumping" is to put products on the market for sale at a price below the actual cost of production. The definition of this unfair and illegal trade needs to be broadened to include all marketing activities that undermine regional self-sufficiency, national sovereignty and local sustainable productivity of the same or similar commodities and services. The fair market price of commodities and services should be reflective of all costs, including social and environmental. On the basis of full cost accountability, more equitable trade policies can be established, and markets encouraged or protected as the case may be. With a firm bioethical standard that considers social and environmental as well as economic factors, there will be incentives to promote the most ecologically appropriate farming methods and choice of crops for domestic use and for export. The scenario of one country or region harming its constituents or its ecology and natural resources by investing in large-scale production of grain, livestock, cotton, or other commodity, and then compounding this harm by "dumping" such produce on the world market and lowering the fair market price will be averted.

The final irony and tragedy of developing countries becoming dependent on imported food commodities and losing their own agricultural self-reliance is the specter of malnutrition and hunger during periods of rapid inflation, when the world market demand and prices for food commodities like chicken and powdered milk suddenly increase. When a country's agriculture collapses, social strife is inevitable, and with political and economic instability, crime and violence and even civil war are likely. The possibility of a recovery of agriculture becomes ever more remote as the poor and hungry try to raise their own food. Irreparable ecological damage to the land and loss of biodiversity are likely consequences of our lacking the right resources, if not also knowledge, especially of sustainable and conservation agriculture.

These socio-economic, environmental and ethical concerns cannot be ignored by the GATT or by the World Trade Organization (WTO). To farm without harm clearly has international ramifications related to equity and world trade. The adoption and multiplication of nonsustainable, intensive livestock and poultry systems by industrialized countries directed toward high-volume production for export, needs to be looked at from an ethical as well as an economic perspective, and constraints applied for the good of all. The same can be said for new genetically-engineered products of agri-biotechnology, like analog cocoa, vanilla, and nut oils, the production of which will harm those countries dependent on raising these products naturally for export revenues, which are themselves needed in part to pay off the interest accrued by too often misguided development loans.

Ethical and moral imperatives notwithstanding, it would be enlightened self-interest for GATT and the WTO to protect and encourage local agricultural self-sufficiency in poorer countries, since the world market will become increasingly dysfunctional and may well collapse if poverty and socio-economic inequities and strife continue to spread under the compounding pressures of population increase and environmental degradation. The application of bioethics to world trade, especially in the agricultural sector, will do much to help every nation and region maximize productivity and minimize adverse environmental and socio-economic consequences, primarily by encouraging mixed farming systems (including agroforestry and aquaculture) that are most appropriate ecologically and culturally for each bio-geographic region.

Harm of Farm Animal Feeds and Wastes

Meat industry defenders counter the argument that importing food for livestock and poultry from the third world contributes to hunger and poverty by insisting that much of this food comes from crop by-products of cash crops grown for export such as sugar cane, molasses, palm kernel cake, cotton oil seed cake, soya bean cake, rice, wheat bran, and rice polishings.[3] In actuality, this market for by-products simply perpetuates unsound agricultural practices in poorer countries, undermines traditional sustainable farming systems and uses up good land that should be used to feed people first.

This food and feed production aspect of animal agriculture, in enabling farmers to import feed for far more animals than the land can sustain from local resources alone, is a major means of support for intensive livestock and poultry production. But it is ethically, economically, and ecologically unacceptable, in part because of the by-product of animal waste that should be returned, but is not and cannot be, to enrich the land in the regions and countries from which the animal feed originated. Such animal waste has become a costly environmental management hazard and is a cardinal indicator of bad farming practices and agricultural policy. Nitrates, phosphates, bacteria, antibiotic and other drug and feed additive residues, such as copper, arsenic, and selenium in farm animal excrement, overload and pollute the environment and food chain.

A related problem is dealing with the enormous volume of another form of animal waste that the meat industry refers to as "animal tankage." The dried and processed residue of animal tankage from rendering plants contains the remains of dead, diseased, and debilitated livestock and poultry, condemned and unusable body parts and even the remains of road-kill and cats and dogs from animal shelters. Slow, low-heat rendering neither sanitizes nor rids animal tankage of potentially harmful organisms, heavy metals, and other hazardous residues. Farm animals, companion animals, and consumers are all put at risk since this by-product of animal agriculture is added to pet foods, livestock and poultry feeds, and is even sold as fertilizer for farm, home, and kitchen gardens. Studies have linked bacterial food poisoning in humans and bovine spongiform encephalopathy in cattle and other animals (that may also be transmissible to humans) with this industry practice of including animal tankage in farm animals' food. Obviously if consumers responded wisely by reducing their consumption of meat and other animal produce, the magnitude of these problems would be significantly reduced with great economic savings.

Finding Solutions

Those who believe that farm animals do not play a vital ecological and economic role in sustainable crop production and range management are as wrong as those who claim that intensive livestock and poultry production are bioethically acceptable because they cause no harm. Now is the time for openness and objectivity and a coming together of all parties and sectors of society involved in the production, marketing and consumption of food. We all need to support the development, adoption and market viability of humane and ecological farming practices that enable farmers to farm with less harm.

The industrial factory system of the animal component of modern agriculture is inhumane and bioethically unacceptable. Making the retail price of meat "cheaper" through tax subsidies and price supports, through better vaccines and biotechnology, through irrigation projects and further deforestation and draining of wetlands, makes it even less bioethically acceptable. So will innovations in meat safety inspection, handling, and processing, including food irradiation, since a full and fair cost accounting will still show that producing meat as a dietary staple causes far too much harm.

The question of the rightness or wrongness of eating meat and of killing animals is not the central issue or primary bioethical concern of the humane movement. Our primary concern is the need to implement less harmful alternatives to contemporary animal agriculture, with its factory farms and feedlots. In the process of producing affordably priced meat as a dietary staple, these intensive livestock and poultry production systems cause harm to farm animals in terms of environmental and production- or husbandry related stress and disease; harm to the agro-ecology, to wildlife, biodiversity, and natural ecosystems; harm to family farms and rural communities; harm to consumers and also to the indigenous peoples of the third world. The antidote is in the adoption and public support of less harmful organic and other alternative, sustainable crop and livestock production practices that are humane and ecologically sound.

The ethic of reverential respect for life and for the land is the guiding bioethical principle of a humane, socially just and sustainable agriculture and society. To question agricultural practices, including new developments in genetic engineering biotechnology that may cause harm, be it to the environment, to sectors of society or to domesticated animals and wildlife, should not be judged as unscientific or obstructive to progress. Surely the essence of progress is to apply science and ethics in the development and adoption of agricultural and other practices and industries that cause the least harm and the greatest good to the entire life community of earth. We cannot sacrifice the good of the environment or of rural communities for the short-term good of the economy, for society will suffer—if not this generation, then the next. Likewise we cannot sacrifice the good of farm animals or of the soil in the name of productivity and labor-substituting technological innovation and marketing, without ultimately harming the economy and the health of the populace.

A Humane Diet

We humans are a highly adaptable primate species. One feature of our adaptive success is our physiological capacity to be omnivores. This capacity to utilize a wide range of food sources, from fruits to nuts and meats to maize, is universal. As I detail in my book *Eating with Conscience: the Bioethics of Food*, for most people around the world, a primarily plant-based diet, with animal products as supplements or condiments, has been shown to be the keystone for a healthy life, economy, and environment (see Fig. 6.1).[4] With rare exception, most people can eat and digest almost anything that other mammalian species can digest (with the notable exception of cellulose), and have developed remarkable ways to preserve and enhance the nutritive value and palatability of a diversity of natural foods. Cultural and ethnic differences in cuisine reflect biogeographic and seasonal variations in food types and availability. This ethnic diversity provides a rich cornucopia of culinary delights and is a source of new crops and food products for an increasingly cosmopolitan marketplace. From this cornucopia, we can select some of the most tried and true diets that have been "human tested" for countless generations, and that are ecologically sound and sustainable. One classic example is what is generically termed "Mediterranean cuisine" that integrates various ethnic foods from this biogeographic region to provide an extremely healthful, relatively low-cost, and ecologically sustainable diet that relies on very little animal protein.

This revisionist view of the USDA's nutrition education program and Eat Right Food Pyramid (which does not, for political reasons, sufficiently encourage a reduction in animal fat and protein consumption) accords with the aims of ethical vegetarians and "conscientious omnivores" to reduce the amounts of animal protein and fat in their diets. An animal-based agriculture and meat-based diet are neither good for the planet nor for one's health. These views, now more widely accepted and promoted by health experts and authorities like the World Health Organization, confirm the connections between a healthful diet and humane and sustainable agriculture. The many benefits of farming without harm and eating with conscience are thereby gaining greater recognition. Such recognition will do much to encourage traditional farming practices and ethnic foods, and prevent the loss of biocultural diversity in world agriculture as well as in the kitchen, which is under siege by the promoters of meat and other animal produce.

Overproduction and overconsumption of food go hand in hand. Meat consumption in the U.S. continues to increase. The average American consumed 204 pounds of meat in 1993, including 65.4 pounds of beef, 52.3 pounds of pork, 68.6 pounds of chicken, and 17.8 pounds of turkey.[5] The relatively low cost of food in the U.S. compared to other countries, coupled with increasingly sedentary life-styles, is responsible for the alarming finding that, according to the National Center for Health Statistics, one in three American adults is now seriously overweight, and the average body weight is increasing.[6] This translates into 58 million people being at

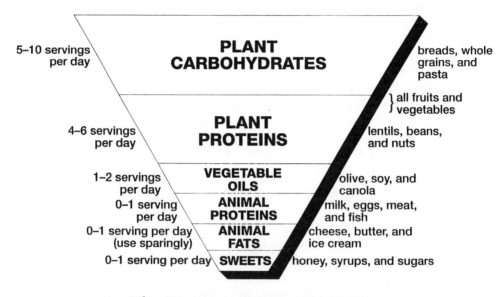

5–10 servings per day	**PLANT CARBOHYDRATES**	breads, whole grains, and pasta
		} all fruits and vegetables
4–6 servings per day	**PLANT PROTEINS**	lentils, beans, and nuts
1–2 servings per day	**VEGETABLE OILS**	olive, soy, and canola
0–1 serving per day	**ANIMAL PROTEINS**	milk, eggs, meat, and fish
0–1 serving per day (use sparingly)	**ANIMAL FATS**	cheese, butter, and ice cream
0–1 serving per day	**SWEETS**	honey, syrups, and sugars

The Humane Dietary Keystone For a Healthy Life and Planet

One serving is equivalent to approximately 1/2 cup cooked rice or pasta; 1/2 cup chopped carrots or broccoli; one slice of bread; one apple; half a grapefruit; 2 tablespoons raisins or nuts; 1/4 cup vegetable oil; 3/4 cup diced lean meat or fish (*meat* includes poultry, beef, pork, and seafoods); 1 cup low-fat milk or yogurt; 1/2 cup low-fat cottage cheese; 1 ounce hard cheese; and 1 tablespoon honey. Fresh, minimally processed, and certified organic foods are preferable, as they are rich in essential trace minerals, vitamins, and fatty acids.

Fig. 6.1. The Humane Dietary Keystone for a Healthy Life and Planet

increased risk for heart disease, diabetes, cancer and other chronic ailments. And overeating has spawned a weight-loss industry that reaps $40 billion per year from American consumers, more than most countries spend on food. Furthermore, the U.S. food industry spends some $36 billion a year on advertising.

The consumer trend in industrial society toward nutritional illiteracy, agricultural amnesia, and culinary catatonia, fostered by the microwaveable frozen meal industry with its prepared and processed convenience foods and relatively meaningless labeling of ingredients and daily recommended allowances, are symptomatic of the disintegration of agriculture and culture. So are the diseases of an overconsumptive and malconsumptive (and malcontent) society that justifies health spas, costly coronary bypasses, and liposuction to remove excess calories. The rest of the human population might well aspire to live this way, but suffers from malnutrition and even starvation due in part to the insatiable appetites of the industrial world.

Farm with Less Harm

Thomas Merton in a manifesto decrying the mistreatment of intensively-raised farm animals wrote:

> Since factory farming exerts a violent and unnatural force upon the living organisms of animals and birds in order to increase production and profits; since it involves callous and cruel exploitation of life, with implicit contempt for nature, I must join in the protest being uttered against it. It does not seem that these methods have any really justifiable purpose, except to increase the quantity of production at the expense of quality—if that can be called a justifiable purpose. However, this is only one aspect of a more general phenomenon: the increasingly destructive and irrational behaviour of technological man. Our society seems to be more and more oriented to overproduction, to waste, and finally to *production for destruction*. Its orientation to global war is the culminating absurdity of its inner logic—or lack of logic. The mistreatment of animals in "intensive husbandry" is, then, part of this larger picture of insensitivity to genuine values and indeed to humanity and life itself—a picture which more and more comes to display the ugly lineaments of what can only be called by its right name: barbarism.[7]

In order to farm with less harm, we must all consume and generally live so as to cause less harm to ourselves and the rest of Earth's Creation. This means reducing the production and consumption of meat in those countries where meat is a dietary staple. It also means refining how animals are raised, transported and slaughtered, and replacing animal protein and fat with cheaper vegetable fats, oils and proteins. Our human population of a soon-to-double 6 billion (of which 1.6 billion are malnourished today) will need to increase the current livestock population of some 4.5 billion to maintain the status quo and public demand for meat. It can do so only if it is prepared to accept a loss of biodiversity and nonrenewable resources, and cope with the attendant environmental and economic risks and costs. We have a better chance of predicting and preventing this, if we begin to incorporate bioethics into the public and corporate policy decision-making process and into our collective vision of what future we are creating this and every day. The first principle of bioethics in agriculture is like the good doctor's Hippocratic aphorism: do no harm.

Protecting Nature and Wildlife

The fact that an agriculture that is based primarily on using good land to raise feed for livestock is nonsustainable is at last being recognized by conventional agriculturalists. The conservative Council for Agricultural Science and Technology published a landmark report in 1994 by agronomist Paul E. Waggoner entitled *How Much Land Can Ten Billion People Spare for Nature?* In his introduction Waggoner writes, "Today farmers feed five to six billion people by cultivating about a tenth of the planet's land. The seemingly irresistible doubling of population and the imperative of producing food will take another tenth of the land, much from Nature, if people keep on eating and farmers keep on farming as they do now. So farmers work at the junction where population, the human condition, and sparing land for Nature meet."[8]

With this premise, and using the latest data from around the world, Waggoner proceeds to show how "smart farmers" can harvest more per plot and thus spare some of today's cropland for Nature—if we help with changed diets, never-ending research, and the encouragement of incentives. Among the points the report makes are:

- Calories and protein equally distributed from present cropland could provide a vegetarian diet for ten billion people.

- The global totals of sun on land, CO_2 in the air, fertilizer, and even water could produce far more food than 10 billion people need.

- By eating different species of crop and more or less vegetarian diets, we can change the number who can be fed from a plot.

- Recent data shows that millions of people do change their diets in response to health, price, and other pressures, and that they are capable of changing their diet even further.

- Given adequate incentives, farmers can use new technologies to increase food productivity and thus keep prices level despite a rising population. Even better use of existing technology can raise current yields.

- Despite recurring problems with water supply and distribution, it is possible to raise more crops with the same volume of water.

- In Europe and the United States, rising income, improving technology, and leveling populations forecast diminishing use of cropland.

The first most important step for every caring person to take is to choose a humane diet. This bioethical imperative should be on top of every nation's agenda, since choosing a humane diet is a vital component in the prevention of animal suffering, in human health-maintenance, and in biodiversity and natural resource preservation. But encouraging people to make this caring and enlightened choice is politically controversial and is still seen as an economic threat by those who have a vested interest in stopping real progress in agriculture.

A revolution in agriculture is gaining momentum nationally and internationally to make it more ecologically sound and environmentally and consumer friendly. The buzzword is sustainable agriculture and our task is to make it humane. With the support of caring consumers, this revolution will succeed, but only if it is made humane. The ultimate goal to farm without harm is attainable, provided those farmers and ranchers and food wholesalers and distributors who care, are supported by all of us choosing a humane diet.

Humane Sustainable Agriculture:
Bioethical Principles and Criteria

Many of the national agribusiness groups who oppose the humane sustainable agriculture movement today will support it tomorrow when there is a clearer understanding of our motives and of the bioethics and profitability of farming nonviolently. Such understanding will lead to a shared vision of a brighter future for all, beyond the short-term goals and imperatives of the world marketplace.

The application of bioethics to evaluate developments and current practices in agriculture will facilitate the adoption of humane practices. There are many academicians, politicians and others who still believe that factory farms and feedlots help America lead the world in producing meat at the lowest cost, and that to abolish this type of farming would hurt the poor who could not afford more humanely and ecologically raised, organically certified meat and poultry. A broader bioethical perspective would enable these groups to see that factory farms and feedlots are neither efficient nor sustainable ways of producing food for human consumption. All new agricultural products, processes and policies should be subject to rigorous bioethical evaluation prior to approval and adoption in order to promote the "farm without harm" ideal, and the goal of sustainability.

When we start from the bioethical principle of farming with the least harm, it is obvious that we should consider the suffering of animals used in farming as suppliers of draft power, fertilizer, fat, fiber, protein, skin and other by-products for human use. If including some animals in the farming system, to make it profitable and sustainable, causes less harm to the ecology of a particular biogeographic region than excluding them, then the humane incorporation of animals should be guaranteed. It is surely the right of animals to be humanely treated, and our obligation. There can be no grounds except sheer greed to justify inhumane husbandry practices. But this is so often done in the name of efficiency and cost-savings. An agriculture that accepts cruel treatment of livestock and poultry is unethical and dysfunctional.

In order to avoid the costs and consequences of intensive animal-based industrial agriculture, we should first and foremost have a soil-based (rather than petrochemical and drug-based) agriculture that utilizes various crops, forages, and animal species sustainably within the limits of available renewable local natural resources, and either enhances or causes no net loss of natural biodiversity. The guiding principle of nonviolent agriculture—to farm without harm—is an ideal that we may never attain, but should not therefore discard. Rather, the degree of humaneness, the quality of life afforded to farm animals, should be the cardinal indicator of sustainable profitability and social acceptability of those farming systems that have integrated animals as essential ecological components. Likewise, maintaining soil and water quality and biodiversity are the basic bioethical criteria for social acceptance of those farming systems that function profitably and sustainably with or without animals. But until national and international bioethical accord and global harmonization of humane farming standards and practices are achieved, great effort will be needed to protect humane and sustainable agricultural systems and communities from unfair competition and possible annihilation by industrial agriculture. The future of agriculture if it is to be sustainable, must be guided not simply by the imperatives of human need and greed, but by the compassionate ethic of ahimsa: of avoiding harm to other living beings, human and non-human; plant and animal, wild and domesticated, either directly, or indirectly as by damaging the environment.

The major criteria for bioethical evaluation are illustrated in Fig. 6.2, which demonstrates the interconnectedness of these interdependent criteria, and that they all converge on economics or full cost accounting. These bioethical criteria include safety and effectiveness; social justice, equity and farm animal well-being; environmental impact, including harm to wildlife, loss of ecosystems and biodiversity; socio-economic and cultural impact, especially harm to established sustainable practices and communities; and accord with established organic and other humane, sustainable agriculture practices, standards and production claims. Farming without harm and choosing a humane diet are coins of the same currency that will forge a strong alliance between urban consumers who care, and rural producers who share the vision of a humane and socially just agriculture and society.

Conventional agriculture is now adopting genetic engineering biotechnology to create new varieties of crops and farm animals. The implications of this new technology in agriculture and medicine, and the need for rigorous applied ethics, as well as scientific assessment of risks and benefits, are discussed in the next chapter.

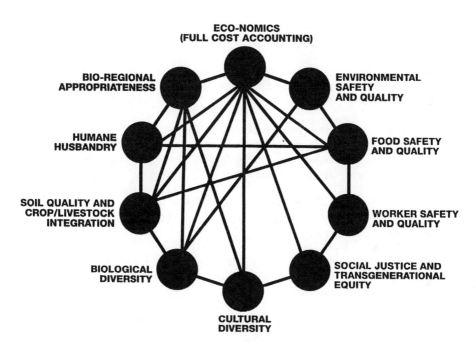

Fig. 6.2. To Farm Without Harm: Bioethical Criteria and Principles of Organic Sustainable Agriculture

POSTSCRIPT

The Ten Commandments of
Humane Sustainable Agriculture:
Farmers', Ranchers', and Consumers' Pledge

1. Make a covenant to transform conventional industrialized agriculture into a profitable and equitable system of food and fiber production that is humane and ecologically sound.

2. Practice and support agricultural methods that are good for all: for farmers, ranchers, the land, animals and consumers.

3. Support local farms and farmers' markets in the community or region that offer produce from organic and other alternative sustainable methods of food production.

4. Purchase no animal produce from factory farms where veal calves, pigs and poultry are raised in complete confinement, with no access to the outdoors. Likewise, do not purchase from factory feedlots where thousands of cows and beef cattle are kept in dirt lots.

5. Support state and federal legislation and civil initiatives to encourage the adoption of humane sustainable agriculture (HSA) and discourage the proliferation of factory farming.

6. Stay informed and involved by joining local and national organizations promoting "eating with conscience" campaigns, vegetarianism, animal rights, and consumer rights.

7. Commit to informing others about why supporting HSA is good for you and good for all.

8. Reach out and encourage local institutions—schools, hospitals, church groups—to engage in community supported agriculture.

9. Appeal to grocery stores and restaurants to provide certified organic foods and more humane animal products, like eggs from uncaged hens, free-range pork and beef, and rBGH hormone-free dairy products.

10. Covenant between urban consumers and rural producers to farm without harm and eat with conscience.

ADDENDUM

Avoiding Nutrient-Deficiency Harm

Health professionals and others are still concerned about the adequacy of vegetarian diets. Sanders and Reddy have shown that a properly selected vegetarian diet can meet all the requirements of adults, and also growing children.[9] Special attention should be focused on ensuring adequate iron, calcium, and vitamin D and B12 intake. Whole-meal bread as a staple (superior to rice) prevents iron deficiency. Vegans (who consume neither eggs nor dairy products) are advised to take vitamin B12 supplements. For strong bones and good teeth, some exposure to sunlight is advisable as one source of vitamin D, plus vitamin D supplements. Calcium supplementation is advisable for those on a macrobiotic diet, or diet that is high in unrefined cereals that contain phytic acid, which interferes with calcium absorption. Vegetarians and vegans are also advised to use oils like soybean and canola instead of sunflower and safflower, which have a low ratio of linoleic to linolenic acid, because a high ratio inhibits the synthesis of docosahexaeonic acid, which plays an important role in brain and ocular functions.

CHAPTER 7

Humane and Bioethical Concerns of Genetic Engineering Biotechnology

Humankind now has power over the atoms of matter and the genes of life. Scientists are now using genetic engineering, which entails manipulation of genetic material in plants, animals and microorganisms, including gene deletion and gene splicing, to introduce genetic material from one species into another. Any new technology brings with it costs and risks, as well as benefits. The benefits of this technology will only be assured if the potential risks and costs are fully acknowledged and, most importantly, if international regulations work to ensure benefits to all rather than to the vested interests of a few.

Bioethicists and others have debated the pros and cons of the human genome project. The risks to animals are considerable. They range from suffering as a consequence of genetic manipulation that results in physical abnormalities, to habitat disruption following the deliberate or accidental release of genetically engineered life forms. The non-medical risks to humans are environmental, social, and ethical. This new technology could be used to screen people and deny some the right and access to equal and fair employment and to medical health and life insurance. For example, women identified as carrying the gene associated with susceptibility to breast cancer may be denied employment because of increased risk of employer liability (as in a chemical factory) and health insurance costs.[1] The technology could also lead to a new era of genetic enhancement where desired traits like increased height, strength, intelligence, disease resistance, and longevity might be accomplished through insertion of genes into developing human embryos. The technology and understanding of gene mechanisms influencing such traits are crude today, but tomorrow could be much improved. Another, perhaps farfetched, spinoff from the human genome project could be the development of new biological weapons, engineered bacteria or viruses, for example, that are harmful to one particular ethnic group of people. An international convention is now being called for to control the proliferation of such "ethnic bombs" after it was revealed that Israeli scientists were involved in such research to target Iraqis and other Arab communities.[2] The moral complexities of this new technology are profound and far-reaching.

Genetic engineering is already a threat to the agricultural economies of some third-world countries. One example is synthetic production of oils from genetically engineered crops, like rape and soya beans, which is in direct competition with export of the natural forms of these products, like palm and coconut oils from Africa and other third-world regions whose economies are dependent upon this export.

Properly applied, genetic engineering could be directed to help improve medical and veterinary diagnostics and therapeutics and to develop vaccines against tropical diseases, and new and safer immunocontraceptives to help enhance the quality of life and security of half the world's population. It could reduce the need for large families, large herds of cattle, and flocks of goats and sheep. The 4 billion animals, along with the 6 billion people on the planet, are now a major threat to the economy, the environment, and biodiversity. Biotechnologies aimed at improved health, self-sufficiency, and life expectancy could help control population in the third world. As livestock become healthier and more productive, fewer will be needed to meet human needs.

Biotechnology should not, however, be seen as a panacea or a substitute for conventional technologies, the most basic of which are good farming practices in accordance with the land ethic and the principles of humane sustainable agriculture.

Current trends in some agricultural applications of this new technology are cause for concern. These include the development of crop varieties that are genetically engineered to be resistant to herbicides, and to produce their own pesticides. Also cause for concern are genetically engineered vaccines and other products, like rBGH, that artificially improve the productivity and adaptability of intensively raised animals, who should instead be raised more humanely and consumed less in developed countries.

The creation and patenting of so-called transgenic mice that carry various defective human genes to serve as models for human diseases may be of very limited medical value. Only a mere two percent of human diseases are caused by single gene defects. A similarly naive reductionistic approach is evident in the human genome project's and similar projects' experiments on farm animals aimed at identifying good and bad genes.[3]

The genetic engineering of plants, algae, and bacteria to produce various valuable materials is well underway. Products range from new vaccines, medically valuable peptide proteins, food additives, enzymes, analogs, and various oils and fuels. Care is needed to assess consumer risks, especially allergic reactions to new food ingredients, and to prevent genetic contamination by such new life forms into the rest of the environment.[4] Examples of harmful consequences include the spread of a new gene by cross-pollination resulting in herbicide resistant "weeds," or the development of new insect pests, plant viruses, or fungus diseases encouraged by newly produced peptides. Birds and insects may be harmed when they eat certain parts of genetically engineered crops like those that produce their own pesticides. They can also be harmed when they eat insects that have been eating these new crops. The ecological repercussions and loss of biodiversity could be considerable. Appropriate containment facilities should be mandated to prevent such harmful repercussions. No technology is risk free. Reasonable risk-control measures need to be implemented. Costs need to be measured against anticipated benefits. The handling of crop and other waste by-products of biotechnology that may be hazardous for use as livestock feed or fertilizer needs particular attention.

Potential risks and unanswered safety questions related to genetically engineered crops and foods are resulting in consumer boycotts and calls for a moratorium on planting all such crops. An estimated 60 percent of all field crops grown in the U.S. in 1999 were transgenic. Many have antibiotic resistance gene markers that could aggravate the problem of bacterial resistance to antibiotics in humans, farm, and other animals. Most genetically engineered crops are infected with viruses since viruses like the cauliflower mosaic virus are spliced with genes and used as "vectors" and "promoters" of those desired genes in the resulting plant. These viruses could recombine with other harmless ones naturally present in the plants' cells to create new and possibly harmful viruses. The unpredictability of introduced genes, the genetic instability and the deleterious pleiotropic effects of introduced genes of genetically engineered plants are also of concern to both farmers and consumers.[5]

Genetic engineering of farm animals to supply human medical needs, like bioengineering pigs to be organ donors, and goats and sheep to produce pharmaceutical products such as Factor 9 for hemophiliacs, raises the medico-ethical issue of human genetic parasitism on other species. Called *molecular* or *pharmaceutical farming*, this new way of exploiting animals to help us treat, but not prevent, our own genetic defects evokes the specter of increasing dependence upon animal exploitation. This is likely to occur, and more animals will be needed, as the incidence of genetic diseases increases in the human population in part as a result of an increasing survival rate.

While the medical benefits of appropriate biotechnology are considerable, with the advent of new diagnostic tools, safer and more effective vaccines and pharmaceuticals, two other concerns need to be addressed: first, the irony of using this technology to bioengineer humankind to adapt to an increasingly polluted and poisoned environment, and second, the suffering of animals genetically engineered to serve as models for human diseases. Many diseases are to a large measure brought on by ourselves—by our dietary habits, by the misuse of agricultural chemicals, and by industrial pollutants that poison the environment and contaminate what we eat and drink. These chemicals can weaken our immune systems, making us more susceptible to AIDS and other future epidemic diseases, and play a role in the development of genetic disorders, mutations, and birth defects. They may also play a role in the etiology of somatic impairments, such as cancer and neurological diseases such as Alzheimers.

Genetic engineering of animals should be questioned if not done primarily for the benefit of animals—or at least with some reciprocal benefit to them. The creation of transgenic animals afflicted with defective human genes that cause birth defects, cancer, brain tumors, and other diseases along with much suffering is ethically and scientifically questionable in terms of progress in human medicine. The first medicine is prevention, and the first preventive measure needed by the industrial world is to clean up the environment and eliminate industrial pollution and agricultural dependence upon pesticides and other agrichemicals, animal drugs, and hormones.

The patenting of genetically engineered animals is to be opposed and a moratorium ought to be imposed by all governments until the above concerns have been fully addressed. It is indeed ironic that the first animal to have been patented was the "onco mouse," a transgenic mouse highly susceptible to breast cancer. The mouse was developed at Harvard University for DuPont Nemours chemical company, which holds the patent. DuPont Nemours is one of several multinational corporations that profit from manufacturing hazardous pesticides and from treating the victims of cancer.

The ultimate concern over the creation of genetically engineered animals and other life forms, and the patenting of life, is the ethical and legal value system of a rising biotechnocracy that continues to sanction animal suffering in the name of medical progress, and unsound agricultural practices on the grounds of efficiency and productivity. Our ethical and legal value system has yet to respond to the above concerns for the ultimate good of all. The democratic, humane, and environmentally safe applications of genetic engineering biotechnology are prerequisites to maximize its benefits and minimize its costs and risks. But such wise applications are undermined by an attitude toward creatures and Creation that sees life as a resource or commodity. And in the process, it reveals a policy of biological fascism, which is unacceptable in any civilized society.

Concerns about Genetically Engineered Foods

The manufacture of genetically engineered foods raises many questions. These ethical questions take us beyond such issues as the acceptability to vegetarians of fruits and vegetables that contain animal genes, and whether or not products of biotechnology qualify as organic if they are raised without pesticides and other agrichemicals. The U.S. government set a precedent in 1994 with its approval of the use of genetically engineered rBGH and sale of milk from treated cows with no mandatory label to inform consumers. By 1999, the Food and Drug Administration (FDA) had declared dozens of engineered crops safe. These include corn that produces its own insecticide (called Bt) and herbicide resistant soybeans, two types of crops that are used widely in processed foods, including baby foods. The FDA continued to stonewall consumer groups' appeals to have all these "Frankenfoods" labeled as having been genetically engineered. Regulators contended that labeling was unnecessary because they see no foreseeable risk to consumers. But in reality manufacturers fear that labeling might turn consumers away, and that would hurt their market.

These genetically engineered foods should be labeled because they are far from natural. When they contain genes from other plants, viruses, bacteria, insects, and various animal species, they don't simply contain inactive alien genes. These genes are manufacturing various chemicals, some of which may be in quantities humans have never eaten before. Allergic and other reactions to such foods are highly probable, and the effects on the developing human fetus and on those with poor immune systems (infants and the elderly) remain to be determined. Shelf life may be extended by slowing down surface spoilage on bioengineered fruits and vegetables, but freshness and related nutrient value may be lost as the time between harvest and consumption is extended.

Other questions need to be asked about failing to label foods that have been subjected to genetic engineering, and of selling them in the first place. Who will be the first to benefit from these new foods? Will food be cheaper and of better quality for consumers? Will these new foods lead to monopoly in the marketplace where, without labeling, consumers will be unable to exercise their right and power of choice? More pointedly, what will their impact be on the environment and agriculture, on how farmers farm and what they farm? What will their contribution be to helping, or aggravating, poverty and world hunger? Billions of dollars are being invested in research, development, and in promoting public acceptance of these new foods. We must ask: do we really need them? Do they have any place in a socially just, ecologically sound and sustainable agriculture?

These questions are rarely asked. But it might be best to consider more than just those questions that concern costs to consumers and potential public health risks, however important they are in the short-term. In the long-term, the costs and consequences of creating genetically engineered grains, fruits, and vegetables may be far more serious than we ever imagined.

Public Involvement in Government and Industry Policy Decision-making

There are considerable gaps in credibility, trust, and understanding between various public interest groups, government agencies, and private, corporate interest groups. There appears to be an alliance between the governmental regulatory sector and the commercial-industrial sector that precludes public input and open dialogue. They seem to have no ear for public concern. The FDA approval of genetically engineered rBGH for use in dairy cows—in total disregard of concerted public opposition and documented scientifically and economically valid concerns—illustrates how strong this political alliance has become. (By way of contrast, rBGH has been prohibited by the Canadian government for use in dairy cows because of animal health and welfare concerns and also by the European Union because of unresolved consumer health concerns.)

Public concern in the U.S., as in many other so-called democracies, is too often dismissed as obstructionist, emotional, and uninformed. Yet government policies made primarily on the basis of "sound science" (i.e., determination of safety and effectiveness) and within the framework of established rule of law and simplistic risk-benefit analyses, preclude many bioethical concerns that the public sector has a democratic right to voice. In reality, the simplistic parameters of profit margins and acceptable risks in a highly competitive world marketplace are the prime determinants of the governmental approval process. For example, the costs of saving an endangered species at risk from land development may be determined by the authorities to be greater than the economic benefits to the local business community of destroying the habitat. Or the costs of cleaning up a toxic dump may be determined to be greater than the scientifically determined health risks to the local community. Approval may be given for a new pesticide, the purported benefits of which outweigh scientifically determined operator and consumer health risks.

Scientific reductionism leads to a fragmented worldview with potentially chaotic and harmful consequences.[6] The corrective is an interdisciplinary *holistic* process of risk/benefit assessments that include determining short and long-term consequences and incorporate a variety of viewpoints rather than just those with narrowly vested interests, such as research and development, marketing and government regulation, and oversight.

The success of American business enterprises in the world market would be greatly enhanced if there were less public opposition that imposes regulatory hurdles and other constraints on innovative technologies. Success is likely to come first to those nation-states that have an educated, informed, and involved populace that participates with government and industry in establishing a bioethical framework around the scientific and legal criteria that are used to deal with the issues that we face as a society today. The European Union's ban on U.S. imports of beef treated with growth stimulating drugs and its moratorium on approving rBGH are interpreted by U.S. business as protectionist interference with free trade. But these are examples of how responsible government responds to public concern, as opposed to simply corporate interests.

In the interim, insofar as agricultural practices and policies, food safety, quality, and security are concerned, the public's right to have all foods labeled to enable selection of organic, humane, natural, local, and non-imported produce should be respected and appropriate action taken. These labeling criteria will help put ethics back into the marketplace and enable consumers to influence the directions industry, government, and society as a whole may take. Such labels will eventually become redundant as the socio-economic system and world marketplace via GATT become ethicized. That such labeling today could, however, be ruled illegal by the World Trade Organization (WTO) is cause for concern.

Public Concerns

An interim report to the USDA Extension Service entitled *Consumer Attitudes About the Use of Biotechnology in Agriculture and Food Production* revealed some significant public perceptions and attitudes.[7] The survey found that:

- Respondents were quite concerned about the socio-economic impacts of biotechnology on farmers and rural communities.

- 85 percent felt it important to label foods if biotechnology was used; 94 percent want to know if pesticides are used, and 88 percent want labels on irradiated foods.

- While some 24 percent of those surveyed felt the use of biotechnology to change plants was morally wrong, 53 percent felt it morally wrong to change animals, and expressed greater concern over eating meat and dairy products developed with biotechnology than they did about genetically engineered fruits and vegetables. (This attitude may soon change, now that leaders of the Christian, Judaic, and Islamic faiths in the U.K. have publicly condoned organ transplants into human patients from genetically engineered pigs.)

- Consumer acceptance of plant-to-plant genetic engineering was 66 percent, falling to 39 percent for animal-to-animal, and 10 percent for human-to-animal transgenic alteration.

- Over half agreed that science and technology have made the world a riskier place and almost half believed that people would be better off if they lived a simpler life without so much technology.

- The most common reason for opposition to biotechnology involved concern that it threatens the balance of nature.
- Over three-quarters of the random population surveyed either agreed or strongly agreed that "the story of creation as told in the Bible is true."

The conflict between Judeo-Christian fundamentalist and secular, materialist attitudes are clearly evident in this public survey. While there was an 80 percent acceptance of animals having rights that people should not violate, almost half believed that "humans were created to rule over the rest of creation." That the orientation of the survey was pro-industry is evident in the statement, "The future of biotechnology in food production is by no means assured without a much more proactive and open (public) educational process."

Corporate Ethics in Biotechnology: Hard Choices, Soft Paths

Corporations like Monsanto and DuPont do not stand alone in believing that the ultimate good for humanity is in sustainable economic development. While critics contend that this is an oxymoron, like military intelligence, the corporate and government consensus that emerged from the Rio Earth Summit is very clear: sustainable development is the top priority for both the developing and industrialized world.

While the definition of sustainable development is being hotly debated, the fact remains that a conceptual seed has been planted—namely, sustainability. This seed could portend the opening of the corporate mind. Before this, economic growth, and ever greater productivity and efficiency were the core values of agribusiness and of the corporate-industrial ethos, with monopoly and patent protection of intellectual property rights providing the necessary security. Without stockholder security and funds from private and public sectors (government tax revenues) to stimulate research and development and the creation of new products, services, and market niches, the spirit of enterprise is likely to be crushed. Economic growth and industrial expansion, however, are no guarantee of sustainability, and in the long-term could be counterproductive.

U.S. based multinational corporations enjoy the security of government and even military protection to ensure that America remains competitive in the global marketplace, and retains jobs for all at home and a quality of life that is the envy of the rest of the world. Through NAFTA, GATT, and the WTO, the influence and interests of multinational corporations will transcend those of nation-states. A new world order based, we hope, on "planetary patriotism," is in the making.

Still embryonic, the corporate seed of sustainability must be nurtured. In the realm of agricultural and industrial biotechnology, the global "greening" or environmental awareness of multinational corporations and of lending institutions, like the World Bank and International Monetary Fund, can become a reality. Enlightened corporate self-interest is linked with sustainable development, just as free and fair trade agreements hold the promise of international peace and cooperation, not competition, conflict, and war. The failure of the petrochemically based Green Revolution is a lesson that corporations, who have not forgotten their history, will not repeat.[8]

Critics of the industrial semioticians who see the changing image of corporations and international banks as Orwellian "newspeak" and "doublethink" rhetoric need to look behind this window dressing. They will find a few naked emperors and some with new clothes, but by and large, when the imperial arrogance of the biotechnocrats is peeled away, and the hyped promises of medical and agricultural miracles through biotechnology scrutinized, the limited, profit-driven horizons of the industrial establishment are clear for all to see. Like any organism, the corporation, regardless of its cultural roots and transnational connections, is as vulnerable as it is opportunistic. Survival and growth, or multiplication, are based upon the ecological and economic principles of sustainability. No organism is secure if it exhausts or poisons its habitat or resource base. The limited worldview of technological determinism and its opportunistic, pioneer ethos is an evolutionary dead end.

In contrast with the negative evolutionary path that some corporations have taken, enlightened corporations with enhanced computer technology and information systems, including satellite monitoring, are following an alternative path of increasing complexity and diversity. The diversification of the petrochemical, pharmaceutical, and food industrial complex into biotechnology is a natural evolutionary sequence following decades of publicly and privately funded research in genetics and molecular biology. Safer alternatives to conventional products and processes that are based upon renewable resources, not nonrenewable resources as is petroleum, are some of the promises of biotechnology.

Safety and effectiveness are not the sole criteria for government approval and public acceptance of new biotechnology products, however. As Greg Simon, chief domestic policy advisor for Vice President Al Gore, has opined, "I predict that if Europeans insist on blocking a safe product" (like bovine growth hormone) "for social and economic reasons, they'll see a flight of capital in biotechnology like they'll never believe."[9] This posture by the Clinton administration, aimed at protecting and promoting U.S.-based technology multinationals, is a continuation of the position of the Bush administration's White House Council on Competitiveness, chaired by Dan Quayle. U.S. biotechnology companies were thrown into disarray when President Bush was asked to sign the international treaty to protect biodiversity. The fear was that the treaty would restrict U.S. companies' access to the plant and animal species that would be protected as intellectual property under the treaty. But this fear eventually dissipated because, as Carl Feldbaum, president of the Biotechnology Industry Organization contended, this industry supports the treaty, because biodiversity is the lifeblood of technology. President Clinton signed the biodiversity treaty in June 1993, but as of 1999 it still has not been ratified by Congress. This support is noteworthy since, as I emphasized in my book *Superpigs and Wondercorn*, a government that is too protective of domestic interests, like the former Bush administration's Competitiveness Council, is likely to do more harm than good to those interests in relation to the global marketplace and volatile nature of venture capitalism.[10]

The sudden turnaround in recognizing the ethical and economic ramifications of the biological diversity treaty is indicative of the internal conflicts of interest and of vision that the biotechnology industry and its cadre of lawyers, scientists, and chief executives had been wrestling with. Having the President of the United States sign the treaty may well signal that the corporate mind is beginning to evolve, realizing that it is enlightened corporate self-interest to explore new avenues of global cooperation. The old paradigm of competition is an anachronism in the evolving new world order, and the U.S. biotechnology industry is still creating its own "Jurassic Park" based on a worldview of economic determinism and monopolistic control of the Earth's genetic resources. In the spring of 1995 I saw the U.S. representative at the United Nations Council on Sustainable Development repeatedly attempt to change the language of an international biodiversity and biotechnology draft convention, essentially to provide a virtually regulation-free framework for the U.S. biotechnology industry. Environmental, social justice, equity, and sustainability issues in the draft were, fortunately, retained.

At the United Nations Biosafety Protocol meeting in Cartagena, Colombia, in February 1999, the life science industry, under the umbrella of the U.S. government and four other countries—Canada, Australia, Chile, and Uruguay—opposed the international treaty supported by 130 other countries including the European Union. They refused to ratify the vital international Biosafety Protocol that grew out of the Convention on Biodiversity reached at the 1992 Earth Summit in Rio de Janeiro, which would have required them to obtain advance approval from importing nations before exporting genetically altered plants, seeds, or other organisms.[11] Ethics is evidently not an issue to those nation-states promoting agricultural biotechnology. They choose to ignore the concerns expressed by the rest of the world about the potential risks and harms of this new technology—risks that an international biosafety treaty on genetically altered seeds and foods might be able to minimize.

The gene rush is as alluring as the earlier gold rush, but the corporate cornucopia of biotechnology products and profits will not be forthcoming if intellectual property rights are not respected internationally. This is the first lesson that this new industry must learn: the genetic resources of nation-states cannot be pirated. Governments must become more responsive to the rights and interests of other countries.

The bioethical complexity of genetic engineering biotechnology is demanding on many levels, from regulatory oversight and compliance to social and economic concerns. As such, it not only challenges the corporate mind in a manner unprecedented by any prior technological innovation, but also can stimulate the development of a broader worldview among corporate leaders that is not as narrowly goal-oriented or directed by short-term profit margins and risk versus benefit ratios. Corporations must consider broader bioethical criteria for product development and evaluation: among others, a product's socio-economic impact at home and abroad, environmental consequences of its introduction, social justice, and transcultural and transgenerational equity issues; and the product's effect on biodiversity, animal well-being and the viability of alternative, more sustainable products, processes and practices.

In sum, this new biotechnology, especially for its advocates and adopters, mandates the development of ecological awareness and is a potent catalyst for it. In *Voice of the Earth*, Theodore Roszak proposes that since the core of the mind is the ecological unconscious, then repression of the ecological unconscious will lead to insanity. This, he argues is the deepest root of collusive madness in industrial society. For Roszak, the path to sanity that involves open access to the ecological unconscious leads us to realize the synergistic inter-play between planetary and personal well-being and to accept that as the needs of the planet are the needs of the person, so the rights of the person are the rights of the planet.[12] Thomas Berry, in his seminal book *The Dream of the Earth*, concludes that human and Earth technologies must be integrated.[13] This ought to be the essence of enlightened corporate behavior and self-interest. In order to accomplish this challenging task, the corporate mind—its ethos and telos—must be changed.

The concept of sustainability that agricultural biotechnocrats are now advocating will find fertile soil, not in the public's accep-tance or in government support of recombinant biotechnology, but in the industry's own creation of an optimal environment to maximize its benefits and profits. The creation of this optimal environment is best assured by subjecting every proposed new biotechnology prod-uct to rigorous evaluation on the basis of the Precautionary Principle and the above bioethical criteria. With this approach, a "soft path" will be laid, as distinct from the "hard path" that products, like rBGH and herbicide resistant seeds, have created for their develop-ers and investors. (The terms *soft* and *hard path*s were first used by advocates of safe and sustainable energy sources in the 1960's, solar and wind-power being soft paths and nuclear reactors epitomized the more dangerous hard path). There was little resistance to genet-ically engineered insulin and "vegetable rennet" because these prod-ucts follow the soft path by meeting the bioethical criteria noted above. The soft path is enlightened corporate self-interest. After the unforeseen hurdles and unprecedented costs that Monsanto Com-pany faced in bringing rBGH to the farmer's gate, the pitfalls of the hard path are clearly evident.

Government approval of rBGH does not mean that the flood gates for new rDNA products and patented varieties of transgenic crops, farm animals and microorganisms have been opened. On the contrary, those start-up biotechnology companies and the big multinational cor-porations that have invested heavily in genetic engineering will both have to follow the soft path. The public outcry and opposition to FDA approval of rBGH and the European Union's ban on this product is a clear signal to the industry to do its homework before investing in re-search and development for a product that will fail to meet some of the bioethical criteria. Safety and effectiveness should not be the sole

criteria for public acceptance of new products. Corporations that do
their homework well will realize the goal of sustainability.

Through genetic engineering biotechnology we have god-like
powers but also choices and responsibilities. The corporate world, in
meeting the challenge of wise and responsible decision-making, will
indeed become "planetary patriots." Evidence of humility, compassion, and ethical sensibility may be lacking behind the veneer of
hyped superpigs, wondercorn, and other touted miracles of bioin-
dustrialism. But as biosophy teaches us that the needs of the planet
are the needs of the people, so the concept of sustainability holds the
realization that the interests of the corporation are the interests of
people and planet alike. Considerable wealth, security, and fulfillment lie in directing this new technology to serve the greater good,
to help heal the Earth and humanity. The alternative hard path will
only quicken the social, ethical, economic, environmental, and physical collapse of our institutions and life support systems.

As I concluded in *Superpigs and Wondercorn*:

> Although "environmentally neutral" biotechnology is a valid
> goal, a better one would be an environmentally and socially
> enhancing biotechnology. This is an attainable goal. It is not
> wishful thinking. Mistakes will be made, but such risks can
> be minimized and this new technology can be applied cre-
> atively and profitably if there is corporate responsibility. And
> this responsibility for the integrity and future of Creation is
> indeed as great as the power we now have over the gene and
> over life itself.[14]

Bioethics of Genetic Engineering

The conservative Hastings Center has published a report that
details the complexity of bioethics, especially in dealing with the creation of genetically engineered animals. This report emphasizes the
difficulties of developing a "grand monistic scheme" that "establishes
a hierarchy of values and obligations under the hegemony of one ultimate value." This approach to dealing with contemporary ethical
concerns is dismissed because while it "may serve the peace of the
soul by reducing internal moral conflict," it would, the authors believe, only work in relatively small and homogeneous communities
and "invariably is bought at the price of the variety and richness of
human experience and significant cultural activity. In this sense it
impoverishes the human soul."[15]

I would argue the contrary. Moral absolutes such as reverence for life and ahimsa can provide both a goal and a common ground for a reasoned and compassionate approach to resolving ethical issues. These absolutes are the cornerstones of a monistic hierarchy of human values that could effectively incorporate the plurality of interests of various segments of society and of different cultures. For example, can we not all agree that to cause otherwise avoidable suffering to fellow creatures is wrong? Instead, this report suggests an alternative strategy: "a coordination of values both *within* and *among* spheres of activity. Contextually coordinating our plural obligations requires a decision making art of moral ecology, judicious weighing of the several obligations in the various contexts at hand be they narrower or wider."[16]

But what catalyst, what shared value or concern is to bring people with differing points of view or value systems together? And against what template is the "judicious weighing" to be done? Surely without the shared goal of enhancing the life and beauty of the Earth, based on the holistic principles of bioethics that extend concern for the good of society to the good of the planet, "the decision making art of moral ecology" will accomplish little. Such a shared goal as enhancing the life and beauty of the Earth is based upon the supreme ethic of reverence for all life and ahimsa. Once such a shared goal is realized and the moral ecology of diverse value systems democratically integrated, via mutual understanding and respect, then a "grand monistic scheme" that "establishes a hierarchy of values and obligations under the hegemony of one ultimate value"—reverence for all life—is possible, if not inevitable. To aim for less is to fall short of realizing the full political and spiritual power of human reason and compassion that the discipline of bioethics embodies. By focusing primarily on the "moral ecology" of diverse and often opposing human interests and values, a human-centered or cosmopolitan rather than eco-centered or cosmocentric worldview and template for public policy will emerge—and we will suffer the gridlock of opposing and unreconciled values in the absence of a monistic unifying principle, such as ahimsa.

Despite its limitations in envisioning the ultimate integration of diverse moral values and human interests into a "grand monistic scheme," this Hastings Report takes a significant steps in this direction, noting that:

> We require systematic ethical responses that genuinely recognize the plural value and ethical dimensions of our

worldly existence. How do we square this circle, which is de-
manded by our overall responsibilities to humans, animals,
and nature? How should such practical decisions be substan-
tively guided? This is an outstanding and unsettled issue. Yet
we may begin to see our way. The first clues come from the
sheer plurality of practices, contexts, values, and obligations
themselves . . . We must become ethically committed, as an
overarching and fundamental moral duty, to this plurality itself: to upholding and promoting the various abiding and culturally significant spheres of human activity amidst the
ecosystemic life and animate world in which they are embedded . . . Ethical atomism or provincialism is practically
impossible and ethically irresponsible. Rather we must *con-
currently* pursue the human, animal, and natural good. First
and foremost we must prevent the significant undermining of
any one domain or sphere of activity, human or natural, for the
sake of others. This involves a mutual commitment, sensitivity, and concern among different human actors with various
contextually defined allegiances. Such coordination requires a
mutual accommodation without forgoing fundamental value
and ethical commitments. We must fashion an ethically and
publicly responsible life that is broadly 'cosmopolitan.'[17]

Such a human-centered or "cosmopolitan" framework for
bioethics is a transitional democratization of ethical values incorporating the moral ecology or plurality of interests and obligations of
contemporary society. It fails, however, to offer any transformative
principle, such as ahimsa, in unconditionally "upholding and promoting the various abiding and culturally significant spheres of human activity," many of which do violence to the sanctity of life in the
name of cultural tradition and social progress.

I find that there is a deeply disturbing kind of rationalism today,
especially amongst those with a scientific background, or who have
an economistic, materialistic, or neo-Darwinian mind-set. One such
person, a top agribusiness corporate executive, told me that farm animal welfare is not a moral or ethical issue. Rather, it's a question of
costs. Another, a senior staff veterinarian with the American
Veterinary Medical Association, dismissed my concerns over the
ramifications of rBGH, asserting that the marketplace is the final
arbiter of ethical issues. "If rBGH causes any problems, then farmers won't buy it," she assured me. The decision to accept or reject

such new products as genetically engineered rBGH cannot be left to a market that is insulated from public censure and manipulated by corporate and government agencies.

Such a laissez faire attitude, professionally and personally, masks a refusal to assume any ethical position or moral responsibility. It is a neutral position often assumed for political reasons, to avoid conflict, or perhaps to avoid exposing conflicts of interest and unethical policies. It is also "scientific" or rational and unbiased to do so because the technocratic ethos, like science and nature, is amoral. But how can society function when we have an amoral economy—that sees farm animal welfare as primarily an issue of financial costs—and government and professional organizations that avoid taking any ethical or moral position, seeing instead the ultimate good as economic growth and market expansion via the proliferation of ever more goods and services?

Can we say that society today is really functional? Leaders speak of the need for more family values, jobs, health care, and education, but to what end if everything that matters is caught in limbo in the moral vacuum of economism and consumerism? Such a vacuum trivializes the human condition and purpose. The despair, alienation, violence, escapisms, and suicides of our youth today are clearly symptoms of our dysfunctional condition. This condition will continue to deteriorate until ethics and morality are woven back into the fabric of the professions and corporate world—and into the World Bank, the White House, NAFTA, and GATT. A laissez faire, market-driven global economy where growth is both a means and an end in itself, will be the nemesis of industrial civilization. What will remain of our sanity, of our humanity, and of the life and beauty of the natural world depends on what we can protect and cherish today, under the broad banner of bioethics and biocultural diversity and accord.

The Naked Emperor of Biotechnology

As in the folk tale of the Naked Emperor, we are shorn of all pretense when we behave as we really feel. For better or for worse, we then show our true natures to the world. Our true natures, from purely ethological and ethnological perspectives, divested of the intellect's defenses and scientific hubris, reveal the truths we live by: the beliefs and values we live for, and also what remains of our feral wisdom and atavistic longings.

The true nature of the life science industry has been revealed by its most naked Emperor, the multinational self-styled life science corporation Monsanto, through Monsanto's acquisition of Delta and Pine Land Corporation, along with the patent rights to the terminator gene technology. This biotechnology, developed by Delta and Pine Land Corporation in collaboration with the U.S. Department of Agriculture (at taxpayers' expense) enables the creation of genetically engineered seeds that farmers cannot save after harvesting to select the best for local conditions, because they will never germinate.

This terminator gene technology, which the U.S. government has patented jointly with Delta and Pine Land (U.S. Patent No. 5,723,765), now owned by Monsanto, is a monopolist's dream. Far from feeding the hungry world, it means that farmers who use the seeds carrying the terminator trait will have to purchase new seeds every planting season. What they will be planting will not be crops that feed the local people first. These crops are biomass commodity crops like wheat, corn, and soy for the food and beverage and livestock feed processors and the myriad subsidiary industries from cosmetics to pet foods.

The risk of the terminator trait being passed on to other crops and to wild plants is very real. But the concerns of plant ecologists and of those many public interest organizations dedicated to a bio-ethically based agriculture and a socially just and sustainable global economy have been discounted and ignored by the life science industry and by the U.S. government. Several other life science industry multinationals, like Zeneca and Novartis, are pursuing similar terminator seed technologies that will require farmers to purchase a costly chemical cocktail to make genetically engineered seeds germinate and develop into plants. And the question remains of whether these seeds are safe.

Exposing the scientific and ethical flaws in a particular field of human activity, like agriculture or medicine not only challenges and tests the truths people live by and their willingness to change.[18] It also threatens a nexus of deeply entrenched financial interests and professional identities. But scientifically and ethically sound and economically viable alternatives could be sought and demonstrated, like sustainable organic agriculture and "holistic" alternative medicines like acupuncture and a healthy diet. Hence, the importance of activist bioethics and the visionary principles of transformative bioethics, in addition to those bioethical principles that fall under the banner of "reformist." Action and vision must be unified if there is to be a reformation in our industrial civilization and in the life science industry and global economy in particular.

Development of the Life Science Industry

The history of the development of the newly self-named life science industry that has come to play a central role in the "Evil Empire" is revealing. (See Fig. 7.1.) During the energy and oil crisis in the 1970s, the major oil companies decided to consolidate and diversify after realizing that the world's oil reserves are finite and increasingly costly to extract. They were fully aware that many of their petrochemical products, used in the Green Revolution to boost world food production (see Addendum), were poisoning the planet, contributing to global warming, and causing cancer and other diseases in wildlife and the people. Small oil and chemical companies were bought out or bankrupted by bigger ones in the industrial "Jurassic Park" of unbridled capitalism and transnational oligopolies.

The oligopolists and their allies in government and academia saw to it that a viable future lay in securing global control not only of natural resources, like oil and uranium, but also of the world market, especially of the food and drug market sectors. They called this monopolistic control "sustainable development." This was promoted through the GATT under the banner of "free trade," and protected by the World Trade Organization and the Trade-Related Intellectual Property Rights (TRIPS) Agreement.[19] In order to gain a monopolistic control of food, the evolving life science industry of petrochemical and pharmaceutical companies bought up independent seed companies, eventually making only their own patented varieties of seeds available to farmers. The fit was perfect when genetically engineered seeds and farm animals were linked productively with the use of herbicides and insecticides, special feed, and production-enhancing drugs for the animals like antibiotics, genetically engineered growth hormone, anabolic steroids, and beta-blockers.[20]

Currently a handful of multinational petrochemical and pharmaceutical companies are using biotechnology to gain monopolistic control of the world market for food and medicines. The kinds of food that are being produced, and the harmful consequences of how they are produced, have yet to be fully accounted for. Concerns range from aquaculture and factory animal farms and feedlots whose animal wastes cause tremendous pollution and animal suffering, and whose food products, contaminated with antibiotic resistant bacteria, create new plagues, to nutrient deficient crops from soils that are chemically sterilized and poisoned with agritoxins. The emphasis on uniform, high volume commodity crop or biomass production has put close to half a million diverse and ecologically more sound

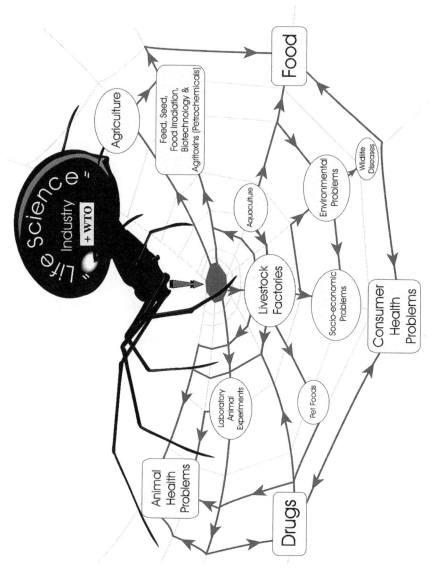

Fig. 7.1. The "web" of the life science industry depicting the linkages between food and pharmaceutical products and the hidden costs and consequences of harming animals, the environment, and the consumer populace.

family farms out of business in the U.S. since the 1970s.[21] The same scenario holds true for other countries where industrial agribusiness has taken hold. The decline in food quality and safety have been of inestimable cost to the public in terms of illness, suffering, time off work, increased insurance costs, government investigations, oversight, and continued subsidies to agribusiness to boost exports.

Norman Myers estimates that the U.S. public subsidizes this toxic food industry to the tune of $69 billion through tax dollars given to agribusiness.[22] Many diagnostic services and drugs that often have harmful and costly side effects would no longer be needed if people consumed organically certified natural foods and beverages in moderation—and people would be much healthier. According to the U.S. Department of Agriculture's Economic Research Service, society now loses up to $7.5 billion yearly in medical costs and lost productivity from bacteria-contaminated meat and poultry products from factory farms. In 1994, researchers at the University of Toronto in Ontario, Canada,[23] estimated that "in 1994 overall 2,216,000 hospitalized patients had serious adverse drug reactions (ADRs) and 106,000 had fatal ADRs, making these reactions between the fourth and sixth leading cause of death." Diet-linked health costs from various forms of cancer and immune system dysfunctions have been estimated to range from $28.6 billion–$61.4 billion per year by the Physicians Committee for Responsible Medicine.[24]

Public health problems from food poisoning to cancer, birth defects, infertility, and immune system dysfunctions, guarantee a growing market for new drugs and diagnostics.[25] It is a conflict of interest for the life science industry to produce more healthful food in less harmful ways, since this would reduce profits in the pharmaceutical and diagnostics sectors. All of this, as well as much animal suffering in biomedical research laboratories, would be greatly diminished if the consumer populace not only supported organic farmers and whole food marketers, but also found doctors for themselves and veterinarians for their animal companions (that sicken and suffer from manufactured pet foods that consist primarily of ingredients condemned for human consumption) who are not prescription pimps for the pharmaceutical industry and who put preventive medicine first. These words may offend some, but I find it infinitely more offensive that animals are being genetically engineered to develop analogs of human diseases and suffer in the process of testing out new drugs.

Currently the life science industry is putting human genes into farm animals whom they call "bioreactors." These transgenic and also cloned "pharm animals" are engineered to produce new pharmaceuticals and other health care products in their milk. Plants are being genetically engineered and patented to not only resist insects by producing their own insecticides, but also to produce "nutriceuticals" such as various amino acids. These will be promoted as new generation "superfoods" that critics call "Frankenfoods." The hoax that natural foods aren't good enough may be pulled off, given the billions of dollars the life science industry is now spending to win public acceptance of genetically engineered foods that the U.S. government, in total disregard of consumer rights, is refusing to label as being genetically engineered.[26]

The science behind this new technology and other life science industry ventures warrant closer ethical evaluation. This is the content of the next chapter. It should not be dismissed as anti-science and anti-progress. Rather, as will be shown, true progress in science and technology cannot be separated from ethics and objective evaluation of motives and consequences.

ADDENDUM

The Failure of the Green Revolution

In a statement by the International Movement for Ecological Agriculture held in Penang, Malaysia, the following comments were made on the *Green Revolution:*[27]

Modern intensive agriculture has conspicuously failed to increase food production and to meet global food and nutrition needs. The claim that the Green Revolution has led to higher crop yields is highly exaggerated and does not reflect a fair and complex comparison with more ecologically sound systems:

- These claims are usually based on the measurement of yield as defined per acre or hectare of land. However, if one takes into account the hidden costs on input subsidies and non-renewable resources, and the costs of ecological damage (leading to lower yields after some time) and furthermore, measure yield against high fertilizer and water costs, then the Green Revolution techniques are highly inefficient. In contrast, the economic soundness of traditional and ecologically better varieties is striking.

- Even more seriously, the Green Revolution measurement of output is flawed because it only accounts for a single crop (eg. rice) and even then only a single component of that crop (eg. grain) whilst neglecting the uses of straw for fodder and fertilizer. Thus it neglects to take into account that there were many other biological resources (eg. other crops, other no-grain uses of the measured crop and fish) within the same land in the traditional system that were reduced or wiped out with the Green Revolution.

- If output is measured in terms of total biomass, a more realistic picture of the performance of the Green Revolution will emerge.

- Although yields of food crops in total have increased, less food is available to local populations. There are several reasons for this:

  - There has been an increase in a few cereals (a large volume of which is fed to cattle in the North) at the expense of pulses and other crops;

  - The increased dependency of Third World farmers and countries on intensive inputs has led to indebtedness and the breakdown of self-sufficiency;

  - Much of the increased food production is exported, thus denying the food to local people;

  - Many areas planted with "high-yielding varieties" (which are actually "high-response varieties" to the applied inputs, including chemical fertilizers and pesticides) are now experiencing diminishing returns;

  - Ecological degradation is leading to reduced yields and to the abandonment of many areas of agricultural land;

  - Losses during storage have increased markedly in many areas;

  - The low prices paid for farm produce and the high prices charged for food in the shops, combined with increased levels of indebtedness, ensure that many farmers cannot afford to buy sufficient food for their families.

CHAPTER 8

Science, Technology, and Ethics

The scientific revolution of the sixteenth and seventeenth centuries laid the conceptual foundations for the industrial revolution. Newtonian mechanical physics and the mechanistic (Cartesian) worldview were illusions created by a reductionistic, rather than holistic, methodology. They offered humanity the false promise of predicting and thus controlling natural processes—and even life itself. Rationalism and empiricism allowed us to know the *how*, rather than the *why*, of things—that which was true synonymous with empirically verifiable facts. The truths of and faith in mechanistic science formed the basis of a new cognitive mode, which enabled people to accumulate capital. As nature became the servant of industrial expansion and material riches, it gave a new sense of security and affirmation to the new faith. Our European forefathers really believed that they could gain control over nature and creation because they had what they thought was the infallible power of objective and instrumental knowledge. Materialistic and mechanistic science became a new faith, and as scientism usurped religion, atheism and human arrogance flourished.

Francis Bacon (1561–1626) is widely regarded as the founding father-philosopher of industrial science and the "Isaiah" of the age of biotechnology. It is through his ideas that the technocracy of today is largely founded, as well as the prevailing attitude toward Nature as a collection of parts, resources, and objects to be exploited and improved upon for the betterment of society. Benjamin Farrington writes that Bacon, "in challenging men—to win power over nature in order to improve the conditions of human life, kindled a new conscience in mankind."[1] Bacon gave religious sanction to man's endeavor to establish and extend the power and dominion of the human race over the universe, by seeing it as a sacred duty to improve and transform the conditions of human existence.

Bacon made the acquisition of knowledge as power (regardless of the means whereby such knowledge was acquired) the supreme ethic. And since the attainment of such knowledge and its application to industrial science is the highest endeavor of humanity, it is of no ethical or moral concern if harm is done to Nature in the process. In other words, Bacon established industrial progress as society's highest principle and priority if not *raison d'être*. Bacon's philosophy was linked with his religious view of dominion as power over the universe and with the Protestant work ethic. His philosophy substituted the notion of spiritual progress and redemption found in the Judeo-Christian tradition for the materialism of industrial progress, which persists up to the present day. By giving industrial progress religious sanction, scientific knowledge became the sacred authority for the new religion of an emerging technocracy.

While in Bacon's time traditional logic was concerned with the philosophy of eternal truths and with the essence of reality, Bacon reviled the lack of practical and material purpose in such thinking. He brought about a reform in logic with his emphasis upon inductive rather than deductive reasoning. And he saw progress in knowledge and progress in power as two aspects of the same process. He established the creed of the modern scientist whose quest for knowledge is the highest virtue since it embodies truth and promises usefulness.

It should not be forgotten that Bacon wrote that "Man is the helper and interpreter of Nature. He can only act and understand insofar as by working upon her or observing her he has come to perceive her order . . . Nature cannot be conquered but by obeying her. Accordingly these twin goals, human science and human power, come in the end to one."[2] While this statement might appear sympathetic toward Nature, it is dubious if Nature needs man as "helper and interpreter." This view planted the idea that it was man's divine role and duty to improve upon Nature. Bacon advised:

> The secret workings of nature do not reveal themselves to one who simply contemplates the natural flow of events. It is when man interferes with nature, vexes nature, tries to make her do what he wants, not what she wants, that he begins to understand how she works and may hope to learn how to control her.[3]

In one of his last publications, Bacon was evidently more reflective and insightful when he wrote:

> Without doubt we are paying for the sins of our first parents and imitating it. They wanted to be like gods; we, their posterity, still more so. We create worlds. We prescribe laws to nature and lord it over her. We want to have all things as suits our fatuity, not as fits the Divine Wisdom, not as they are found in nature. We impose the seal of our image on the creatures and works of God, we do not diligently seek to discover the seal of God on things. Therefore not undeservedly have we again fallen from our dominion over the creation; and, though after the Fall of man some dominion over rebellious nature still remained—to the extent at least that it could be subdued and controlled by true and solid arts—even that we have for the most part forfeited by our pride, because we wanted to be like gods and follow the dictates of our own reason.[4]

This statement is relevant today, in that we have broken "the seal of God on things," in atomic and genetic research especially, and now superimpose "the seal of our image on the creatures and works of God."[5]

There is something distinctly disturbing about Bacon's words, aside from his loathing of Greek philosophy, particularly Plato and Aristotle. While he may be rightly regarded historically as the prophet and herald of industrial science, his visionary advice failed to temper human arrogance and greed, in spite of, if not because of, his invocation of divinity. Philosopher Brian Klug writes that Bacon's talk of obeying Nature rings rather hollow when we recall Bacon's "main objective to make Nature serve the business and conveniences of man."[6] Klug concludes that Bacon clearly invoked God's will and divine sanction to justify the scientific inquisition of Nature.

The use of animals in biomedical research today is presaged in Bacon's utopian saga *New Atlantis*, the protagonist Father of Salomon's House proclaiming, "we also have parks and enclosures of all sorts of beasts and birds, which we use not only for view or rareness, but likewise for dissection and trials, that thereby we may take light what may be wrought upon the body of man . . . We try all poisons and other medicines upon them as well as of chirurgery [sic] as physic."[7] Furthermore, the size, shape, color, behavior, and reproduction of animals were altered in his story and new species were created, a science fiction vision which, through genetic engineering, is a reality today. If anything, Bacon intensified the duality of nature and divinity; in one breath he castigated man for wanting to be like gods, and in the next breath he sought to attain God-like dominion over "rebellious Nature."

According to John F. Haught, Bacon's philosophy of scientific materialism "reduces life and mind to the level of the lifeless and mindless." He goes on to say that it is also "essentially incompatible with a sound environmental outlook . . . A philosophy that theoretically resolves the animate into the inanimate can hardly function as the foundation for policy that strives to prevent this destruction from happening in practice."[8]

The Ends that Science Serves

Science has as its faith a subjective, emotional basis, and this is the love of truth that is shared with all religions. But now there is conflict between science and ethics and religion where there was once none, because science is increasingly used for power and control over life in the service of purely selfish and materialistic ends.

Richard Levins and Richard Lewontin, in *The Dialectical Biologist*, trace many of the negative social consequences of scientific materialism to the use of science to serve capitalist values and goals. They state:

> [The commoditization of science] stands between the powerful insights of science and corresponding advances in human welfare, often producing results that contradict the stated purposes. The continuation of hunger in the modern world is not the result of an intractable problem thwarting our best efforts to feed people. Rather, agriculture in the capitalist world is directly concerned with profit and only indirectly concerned with feeding people. Similarly, the organization of health care is directly an economic enterprise and only secondarily influenced by people's health needs.[9]

Now genetic engineering is being heralded as the latest panacea. It is believed that all will be put right through this new biotechnology. We will have better medicines and diagnostics, "safer" substitutes for chemical pesticides, bacteria to eat up industrial pollutants, and be able to create more productive animals and crops that are even salt and drought resistant. But will people be healthier and have more food than ever before?

The greatest danger of genetic engineering is our lack of fear. Most scientists, industrialists, politicians, and investors believe that our power over the gene is the key to utopia. But the prevailing attitude of the technocracy toward Nature and living beings will guarantee that this new biotechnology will lead to Nature's end, if there is not a radical and enlightened change in attitude and perception.

In *Freedom in the Modern World*, John MacMurray states that "we cannot put our trust in science, for a very simple reason . . . What we do with knowledge that science creates is not the business of science. Science has nothing to do with good or evil, with the satisfaction of human desires . . . And so we have to decide for science what is worth doing before we use science to do it."[10] MacMurray contends that science is knowledge, and that knowledge is power. Since power is the servant of those who have it, science, therefore, is necessarily a servant; it cannot be a master. But it would seem that since we have put our faith in objective knowledge, we have become servants of the power that science gives us over the material world. By making science the religious faith of the modern world, we have lost our souls in exchanging spiritual values and ethics for material knowledge, global power, and control.

In his book *The Reenchantment of the World*, Morris Berman concludes, "Our loss of meaning in an ultimate philosophical or religious sense—the split between fact and value which characterizes the modern age—is rooted in the Scientific Revolution of the sixteenth and seventeenth centuries."[11] He shows that for 99 percent of human history the world was unchanged and man saw himself as an integral part of Creation. He observes that "the complete reversal of this perception in a mere four hundred years or so has destroyed the continuity of human experience and the integrity of the human psyche. It has very nearly wrecked the planet as well."[12]

The Cartesian split between mind and body, subject and object, spirit and matter, is the basis of rational egotism: reality is objectified when we conceptualize the world not as a unified field but as made up of individuated parts. From a more holistic perspective, this worldview is schizoid because in turning relationships and potentialities into objectified parts, we destroy what is ecologically an indivisible whole. Through this, we destroy our "sacred connectedness."

Science and the Technocracy

Philosopher Alan Drengson defines technocracy as:

> . . . the systematic application of technology to all levels of
> human activity, including governmental and economic poli-
> cies which have growth as their central aim. Such growth in
> the contemporary West is often promoted by means of poli-
> cies which favor complex, high technologies. The scale in-
> volved in applying new technologies dictates a need for
> government and corporate planning; thus, only specialists
> can write policy. The aim becomes the control of life by
> means of management techniques that govern the applica-
> tion of the hardware and processes integral to technology.
> Science is narrowed to its less theoretic activities and the
> principal emphasis is upon prediction, control, and applied
> science. The sciences so stressed are thought to be value-
> free. The aim is to reduce all phenomena to those features
> which can be quantified, controlled, and observed directly
> with the instruments produced by technology. . . .

Technocracy has become incompatible with democracy. It is
ultimately destructive because the necessary democratic controls
by government, such as preventing industrial pollution, are incompatible with the principle of "free enterprise." This is aggravated by the fact that short-term human needs and benefits, such
as more jobs, goods, and services, take precedence over long-term
needs and benefits such as clean air and water. The world and
other living beings are only seen in terms of human utility, and
such objectification leads to the nihilism of a world devoid of inherent value.

Drengson continues:

> [. . . Thus,] the technocratic "machine" drives to manage all
> aspects of natural, industrial, and social processes by means
> of centralization, substituting where possible machines for
> humans, rules and laws for morality, social system for com-
> munity, monoculture for diversity, and so on. This drive is
> found in capitalist and socialist nations alike, for the mech-
> anistic paradigm is essentially global, transpolitical, trans-
> ideological, and is closely connected with modern industrial
> technology and its specialized disciplines . . . Humanism, as

homocentrism, joined with the technocratic paradigm, must finally assume the overwhelming responsibility for running everything.[13]

Clearly, there is less and less freedom in the modern world as we get tangled in our own controls.

Religion and Ethics

The technocratic worldview holds firmly to the belief that the problems arising from technologies are consequences of a lack of instrumental knowledge, and that the solution is more research. This worldview doesn't see a need for ethical and spiritual sensibility and empathetic understanding. Buddhist Tarthan Tulka aptly states:

> The disharmony in our lives is reflected in our attempt to control the environment and the conditions which result. We so often attempt to use the world around us for our personal uses and immediate comfort without taking into consideration any wider perspective. In implementing such narrowly-construed purposes, we create imbalances that bring forth both present and future problems for ourselves and others . . . We should now consciously consider our responsibility to restore a balance, an integration of material (scientific and technological) advances with the deeper values of humanity. And when there is a balance of the two ways of thinking—technology can be utilized as a very valuable and creative force.[14]

It is unfortunate that organized religion today is essentially human-centered, emphasizing salvation from the world rather than creative participation in the stewardship of Earth. Most religions are not theocentric, or ecocentric, because they exclude concern for the whole of God's Creation, a concern that must embrace animals and Nature for it to be fully theocentric. The humanocentrism, especially of the Judeo-Christian tradition, has been infused with the ideals of technocracy. The technocratic ideals of industrial growth and the myth of progress toward some future utopian materialist paradise have now gained the power of religious doctrine.

But this worldview is now being challenged. In December of 1987, Pope John Paul II's encyclical *On Social Concerns* was issued. It is the first encyclical to address ecological concerns and gives affirmation and support to the cause of animal rights. It is a direct challenge to the industrial technocracy and a call to accountability, especially for those industries that gain financially from exploiting "animals, plants, the natural elements." The following excerpts from this encyclical are evidence of a more Creation-centered theology:

> Para. 33. True development, in keeping with the specific needs of the human being—man or woman, child, adult or old person—implies, especially for those who actively share in this process and are responsible for it, a lively awareness of the value of the rights of all and of each person. It likewise implies a lively awareness of the need to respect the right of every individual to the full use of the benefits offered by science and technology . . .

> Para. 34. Nor can the moral character of development exclude respect for the beings which constitute the natural world, which the ancient Greeks—alluding precisely to the order which distinguishes it—called the "cosmos." Such realities also demand respect, by virtue of a threefold consideration which it is useful to reflect upon carefully.

> The first consideration is the appropriateness of acquiring a growing awareness of the fact that one cannot use with impunity the different categories of beings, whether living or inanimate—animals, plants, the natural elements—simply as one wishes, according to one's own economic needs. On the contrary, one must take into account the nature of each being and of its mutual connection in an ordered system which is precisely the "cosmos."[15]

There is a dire need today for higher values and an ethical sensibility above and beyond those of material progress. Since the time of Francis Bacon, scientific serendipity has been linked with imperial power and industrial wealth. In *Man, The Bridge Between Two Worlds*, Frantz E. Winkler says:

> If you consider nature as a mere tool to serve man, science will become an instrument of destruction and thus a hazard to the future of man and his world. To the medieval sage science was religion, and its deeper purpose was to serve God in nature. . . . Nature is [now] the main object of scientific exploration. She may well fall victim to analytical urges gone

berserk, unless these urges are balanced by the compre-hending wisdom of selfless love to all her kingdoms.[16]

In the same way, René Dubos, in *A God Within*, cautions:

A relationship to the earth based only on its use for economic enrichment is bound to result not only in its degradation but also in the devaluation of human life. This is a perversion which, if not corrected, will become a fatal disease of techno-logical societies.[17]

Albert Einstein, whose scientific work contributed so much to the development of our technological society, was likewise aware of values far beyond those to which his discoveries have been applied, as in the development and use of thermonuclear weapons. In 1954 Einstein contended that:

The most beautiful experience we can have is the mysteri-ous. It is the fundamental emotion which stands at the cra-dle of true art and true science. Whoever does not know it and can no longer wonder, no longer marvel, is as good as dead, and his eyes are dimmed. It was the experience of mys-tery—even if mixed with fear—that engendered religion . . . I am satisfied with the mystery of the eternity of life and with the awareness and a glimpse of the marvelous struc-ture of the existing world, together with the devoted striving to comprehend a portion, be it ever so tiny, of the Reason that manifests itself in nature . . . The man who regards his own life and that of his fellow creatures as meaningless is not merely unhappy but hardly fit for life.[18]

An ethical and spiritual relationship with others, as well as with the world, seems to have little relevance today. There are many definitions of spiritual—among them, "relating to, or coming from the intellectual and higher endowments or the mind," and "of or re-lating to moral feelings." In *The Teaching of Reverence for Life*, Albert L. Schweitzer says that spirituality is humankind's most valuable tool for progress, arguing:

Three kinds of progress are significant for culture. Progress in knowledge and technology, progress in the socialization of man, progress in spirituality. The last is the most important.[19]

The Roots of Imperialism and Morality

We should consider the historical influence of Roman imperialism upon our contemporary culture's attitudes towards ethics, science, and religion. As John MacMurray emphasizes in *Freedom in the Modern World*, . . . the Roman element in our culture is to be found in our moral tradition. Our traditional morality is a morality of organization, which MacMurray sees as obedience to moral laws which have a social reference. MacMurray suggests that the dominant influence in our civilization is not Christianity, but the pervasive influence of the Roman Empire:

> To this day our culture has remained in that Roman mould. It is essentially imperialist; that is to say, its governing ideal is the maintenance and perfecting of an efficient organization of social life, depending on law, industrial management and the maintenance of power for the defense of law and property. Art and religion have been harnessed to the service of this ideal of administrative organizing efficiency and subordinated to it.[20]

What is particularly stunning in MacMurray's historical treatise is his convincing conclusion that the morality of today is derived not so much from Christianity, as it is from the Roman Empire's stoicism. The ideal of this legacy is the supremacy of reason over emotion, a life of law, principle and rational organization as its goal. The emotions, judged as irrational and thus a potential source of disorder and evil, were to be controlled and suppressed, as was human freedom and the life of the spirit. A person must act rationally, not emotionally, and always do what ought to be done rather than what is desired. Suppression of emotion and personal freedom was a necessary part of the organization and legal administration of the Empire. But MacMurray insists that legal rationalism must be the servant of personal freedom, and that life should be based upon an emotional principle, not on an intellectual one.

The Roman tradition provided the perfect foundation for the later age of reason, the industrial revolution, and the Protestant and Marxist work ethic. The work ethic—whether Protestant or Marxist—embraces a principle that accepts human industry as a virtue. To engage in labor for the benefit of society is a moral duty or obligation. This work ethic is linked with the belief that it is virtuous to exploit and to *improve* upon Nature for industrial purposes, and it further holds that there is nothing morally wrong in utilizing animals for various industrial and commercial purposes for the benefit of society. Thus, to speak out against the destructive industrialization of Nature and cruel exploitation of animals is seen as a threat to capitalism and the work ethic. To suggest that animals have rights that should not be violated without question for the purported benefit of society could be construed as placing animals before, if not above, people who, by following the work ethic, make certain sacrifices themselves for the benefit of society.

Animal rights philosophy does have a libertarian ethos that could be construed as an attack upon both capitalism and the work ethic. The violation of animal rights and of those who serve the industrial technocracies of Protestant and Marxist cultures constitutes an erosion of the spiritual and egalitarian ethics of both Christianity and socialism. John MacMurray writes:

> All organization is a means to an end, and it is to be judged by its results. It is good so far as it makes human life freer and richer. It is an evil thing when it makes life more of a slavery and emptier. And it seems to me that it has in fact done that for us. I do not believe that the inner quality and richness of the lives of the mass of men and women has been made greater and better by the progress of social organization in the last century. It seems to me to be definitely impoverished. Therefore I think that it is time to call a halt and ask ourselves in what the inner significance and value of human life consists. If we can be better human beings by being poorer, then by all means let us get rid of some of our wealth.[21]

That technology has become a means without end intensifies the meaninglessness of materialism. As John Grim states, "If progress is the myth of technology. then mass production is its ritual embodiment."[22] Meaning and significance in life arise from the numinous reality of the phenomenal world—not in the proficiency of our work and efficiency of industry and society.

Science and Religion Unified

Pope John Paul II, at the Einstein session of the Pontifical Academy of Sciences, made the following pertinent remarks vis-à-vis science and religion:

> Applied science must be allied with conscience so that through the triad, science-technology-conscience, the true good of humanity will be served.

Unfortunately, as I had occasion to say in my encyclical *Redemptor Hominis*, "Man today seems always menaced by what he produces . . . This seems to constitute the principal act of the drama of human existence today." (No. 15) Man must emerge victorious from this drama, which threatens to degenerate into tragedy, and he must rediscover this authentic kingship over the world and his full dominion over the things he produces. Today, as I wrote in the same encyclical, "the fundamental meaning of this 'kingship' and of this 'dominion' of man over the visible world, which is given him as a task by the Creator, consists in the priority of ethics over technology, and preeminence of people over things, and the superiority of spirit over matter." (No. 16).[23]

According to the church of Rome, science, through the knowledge gained in its name, can bring us closer to understanding God through Nature, provided, as the council reminds us, there is an attitude of humility. Conversely, Einstein believed in a "God who reveals himself in the harmony of all things, and not in a God who is interested in the actions and destiny of each individual." He believed that science without religion is lame, and religion without science is blind.

Martin Luther King recognized these problems when he wrote:

Science investigates, religion interprets. Science gives man knowledge which is power; religion gives man wisdom which is control. Science deals mainly with facts; religion deals mainly with values. The two are not rivals. They are complementary. Science keeps religion from sinking into the valley of crippling irrationalism and paralyzing obscurantism. Religion prevents science from falling into the marsh of obsolete materialism and moral nihilism . . .

We must work passionately and indefatigably to bridge the gulf between our scientific progress and our moral progress. One of the great problems of mankind is that we suffer from a poverty of the spirit which stands in glaring contrast to our scientific and technological abundance. The richer we have become materially, the poorer we have become morally and spiritually . . .[24]

C. P. Snow, physicist and philosopher, thought that science and religion were like two separate cultures. He emphasized that the "gulf of mutual incomprehension" between intellectuals and scientists sometimes leads to hostility and dislike, particularly among the young of our culture.[25] Some scientists argued that one of the two cultures was better than the other, failing in the process to reconcile the dialectic by realizing that both worldviews are complementary and need to be integrated.

Although Nobel physicist Francis Crick, co-discoverer of the double-helix configuration of the DNA molecule, acknowledged Snow's recognition of the different cultures or worldviews, he wrote:

> The mistake [Snow] made, in my view, was to underestimate the difference between them. The old, or literary culture, which was based originally on Christian values, is clearly dying, whereas the new culture, the scientific one, based on scientific values, is still in an early stage of development, although it is growing with great rapidity. It is not possible to see one's way clearly in the modern world unless one grasps this division between these two cultures and the fact that one is slowly dying and the other, although primitive, is bursting into life.[26]

Contrast this enchanted view of scientism with Wendell Berry's poignant vision of how our technocratic society has evolved, and how destructive it has become:

> Modern urban-industrial society is based on a series of radical disconnections between body and soul, husband and wife, marriage and community, community and the earth. At each of these points of disconnection the collaboration of corporation, government, and experts sets up a profit-making enterprise that results in the further dismemberment and impoverishment of the Creation.

> Together, those disconnections add up to a condition of critical ill health, which we suffer in common—not just with each other, but with all other creatures. Our economy is based upon this disease. Its aim is to separate us as far as possible from the sources of life (material, social, and spiritual), to put these sources under the control of corporations and specialized professionals, and to sell them to us at the highest profit.

It fragments the Creation and sets the fragments into conflict with one another. For the relief of the suffering that comes of this fragmentation and conflict, our economy proposes, not health, but vest "cures" that further centralize power and increase profits: wars, wars on crime, wars on poverty, national schemes of medical aid, insurance, immunization, further industrial and economic "growth," etc.; and these, of course, are followed by more regulatory laws and agencies to see that our health is protected, our freedom preserved, and our money well spent. Although there may be some "good intention" in this, there is little honesty and no hope. Only by restoring the broken connections can we be healed. Connection is health.[27]

Historian Lynn White has made the provocative observation that:

Christianity in absolute contrast to ancient paganism . . . not only established a dualism of man and Nature but also insisted that it is God's will that man exploit Nature for his proper ends . . . In antiquity every tree, every spring, every stream, every hill had its own *genius loci*, its guardian spirit . . . By destroying pagan animism, Christianity made it possible to exploit Nature in a mood of indifference to the feeling of natural objects.[28]

I do not entirely agree with White's assertions, however.

A tragic outcome of dualism is the separation of humanity from Nature and animals, and of religion from everyday life—of the sacred from the secular.[29] When this splitting occurs, it is a symptom, if not a cause, of the inevitable disintegration of society. As the highest ethical and spiritual values become less and less an integral part of daily life, the organization and coherence of society becomes increasingly dysfunctional. Moralistic, legalistic, and punitive correctives become less and less effective as this process of disintegration progresses, yet ironically there is increasing reliance upon them. More laws and lawyers, bureaucracies and regulators, give a sense of order and civility to a civilization in chaos. The only corrective is a return to what some call traditional values, as distinct from the predominantly shallow utilitarian and materialistic values that characterize contemporary secular technocracies.[30] Such traditional, ethical, and spiritual values as respect for the environment and other living creatures, and reverence for the integrity and future of

Creation are barely evident in society today where technological developments and industrial growth have taken place for decades. Whitehall N. Perry[31] in his massive review of traditional wisdom concludes that a secular society that is dedicated to material benefits can never achieve the quality of life for its members as can traditional societies primarily oriented to spiritual ends.

Technology—Control and Responsibility

A *leaderless* technology is one that is not guided and constrained *at its inception or in its conception* by either ethics or traditional values. Instead it is regulated, *post hoc*, by various laws and governmental regulatory agencies. Such is the situation today, but it may change as industry, government, and public interest groups begin to wrestle with the long-term ethical, social, and environmental consequences of the latest and greatest technological development—genetic engineering technology. Indeed, the greatest benefit of this new technology could well be in its forcing us to return to traditional values. And if we are to play God, we should do no less. Otherwise our own nemesis will be assured, and accelerated by the inevitable misapplication of genetic engineering biotechnology.

In *Steps to an Ecology of Mind*, the late Greg Bateson advises:

> We would do well to hold back our eagerness to control that world which we so imperfectly understand. The fact of our imperfect understanding should not be allowed to feed our anxiety and so increase the need to control. Rather, our studies could be inspired by a more ancient, but today less honored, motive: a curiosity about the world of which we are a part. The rewards of such work are not power but beauty.[32]

As our sense of beauty increases, so does our ability to empathize, as does our awareness of loss and of the ugliness we create. Aesthetics can be our salvation, especially when beauty is the apperception of love.

Because Western civilization has become so unbalanced technologically, materially, ecologically, and spiritually, people are discovering that the very goals and related values and ethics of the technocracy which they support, and upon which they have become so dependent, lack personal significance. The rewards by and large do not enrich the inner life of the person. They do not lead to self-

realization. They lack spiritual purpose and meaning. In his book
The Domination of Nature, philosopher William Leiss writes:

> The idea of the mastery of nature must be reinterpreted in
> such a way that its principal focus is *ethical or moral de-*
> *velopment* rather than scientific and technological innova-
> tion. In this perspective progress in the mastery of nature
> will be at the same time progress in the liberation of na-
> ture. The latter, like the former, is a rational idea, a con-
> cept, and achievement of human thought; therefore the
> reversal or transformation which is intended in the transi-
> tion from mastery to liberation concerns the gradual self-
> understanding and self-disciplining of *human* nature. As a
> rational idea "liberation" can apply only to the work of con-
> sciousness, to human consciousness as an aspect of nature,
> and not to "nature" as a totality. The task of mastering na-
> ture ought to be understood as a matter of bringing under
> control the irrational and destructive aspects of human de-
> sires. Success in this endeavor would be the liberation of
> nature—that is, the liberation of human nature: a human
> species free to enjoy in peace the fruits of its productive
> intelligence.[33]

In other words, we must apply our knowledge, or power over Nature,
with wisdom to further our moral and ethical (spiritual) development through focused understanding of human nature. We cannot
master Nature without mastering our own human nature. The ethical and spiritual principles of religion are the truths that we can best
live by. We must use them to help guide and constrain the truths of
science that give us the power of instrumental knowledge, for such
power may then be used empathetically and creatively, not fearfully,
selfishly, or destructively.

The feelings that underlie the motives of scientific inquiry must
not be dismissed. Ethical feeling must temper science and its inquiries
and techniques. By avoiding our feelings, the means whereby certain
goals are attained (such as those that entail the suffering of laboratory
animals, and even human guinea pigs in Nazi concentration camp
laboratories) become divorced from ethical sensibility, which is
founded not upon instrumental accomplishment but upon emotional
sensitivity. The end result, when ethical values are subsumed by materialistic, economic, or other values that may benefit private industry
or society, is that unethical means are sanctioned. Yet in the long-

term, no society can benefit from the suppression of emotional sensitivity, ethical sensibility, and responsibility. This is one of the major dilemmas of modern society, where such scientific-technological advances as amniocentesis, genetic engineering, and the ability to keep brain-dead people alive can lead not only to the abuse of power, but to certain ends that we are not yet equipped to deal with emotionally or ethically. Hence, the need for greater emotional involvement in our actions and aspirations, in our being and becoming. By relying exclusively upon the acquisition of knowledge for power and control over life, and not to clarify our values and motivations, life itself can become merely a means to meaningless ends—meaningless because of the absence of ethical responsibility and foresight.

The means whereby knowledge is gained should be given the same emotional and ethical consideration as are the ends or goals. Without thoughtful consideration, the upshot is unnecessary suffering and unethical exploitation of animals and people. And if the ends themselves are valued in and for themselves only, the result may be inappropriate and even harmful application of such knowledge, as with the use of antibiotics to promote farm animal growth and atomic power to construct thermonuclear weapons. In sum, scientific *freedom*, like personal freedom, is inseparable from emotional sensitivity, ethical sensibility, and responsibility. In the absence of enlightened constraints, the need for costly and often ineffectual legal and governmental controls arises.

When our values are not based on the "virtues" of bioethics—like reverential respect of other life forms, humility and compassion—and on such social virtues as self-respect, self-reliance, personal integrity, and common decency, they become relative and subjective. As Gertrude Himmelfarb so clearly puts it:

> Values, as we now understand that word, do not have to be virtues; they can be belief, opinions, attitudes, feelings, habits, conventions, preferences, prejudices, even idiosyncrasies—whatever any individual, group or society happens to value, at any time, for any reason. One cannot say of virtues, as one can of values, that anyone's virtues are as good as anyone else's, or that everyone has a right to his own virtues. Only values can lay that claim to moral equality and neutrality. This impartial, "nonjudgemental," as we now say, sense of values—values as "value-free"—is now so firmly entrenched in the popular vocabulary and sensibility that one can hardly imagine a time without it.[34]

The nonjudgmental *"value-free"* sense of values that she so clearly describes is an aspect of the scientific method where a subjective consensus is regarded as truly objective because it is held to be value-free. This "doublethinking" and "newspeak" of science, as George Orwell termed the process of manipulating truth to serve the ends of the technocracy, has become incorporated into public policy under the guise of *science-based* analysis and decision making.

Adolph Hitler's scientifically rationalized extermination of Jews, in order to create a genetically *pure* Aryan nation, illustrates the destructive and inhumane consequences of making a "faith" out of the "truths" of science, and of using such "truths" to serve political or other self-serving ends. Science cannot be a substitute for ethics and religious faith. The instrumental rationalism of science that lacks any ethical basis becomes purely self-serving power. No species or culture can be advanced or enhanced by such a worldview. Thus we must question today the politically sanctified scientific manipulation of Nature and of living beings—especially genetic engineering which is now a major *growth industry*—and the unremitting torment of animals in laboratories that are made to suffer mainly because of our own *sins*.

Geneticist Dr. Mae-Wan Ho sees that "genetic engineering biotechnology is an unprecedented alliance between bad science and big business, which will spell the end of humanity as we know it, and of the world at large."[35] Pope John Paul II in his 1998 encyclical *Fides et Ratio* (Faith and Reason) expresses this same concern that "if science does not move beyond the utilitarian it could soon prove inhuman and even become the potential destroyer of the human race."

In pointing out the existence of a biotechnocratic "Evil Empire," I run the risk of being misjudged as an anti-establishment neo-Luddite, an alarmist who fabricates conspiracy theories with malevolent intent. On the contrary, I regard all technologies as being neutral, the good or harm derived therefrom being determined by how and to what ends they are applied. Furthermore, I do not believe that the creators and participants of the Evil Empire have any evil intentions. Rather, without any bioethical principles and constraints, evil consequences can arise from good intentions. We must therefore extend the aphorism that "evil flourishes where good men do nothing," to where good people act without bioethical sensibility. Evil arises when good people rationalize harmful means for beneficial ends and choose to live in denial of the harmful consequences of their values and actions in order to "save face," or for other reasons. Adams and Balfour in their book *Unmasking Administrative Evil* emphasize how ordinary people can engage in acts of evil without being aware that they are doing anything wrong.[36] They manifest "moral inversion" when evil activity is seen as good.

In conclusion, a new paradigm for science, especially in its applications to agriculture, and medicine is urgently needed. This can be attained by an awakening of ethical sensibility, and a more holistic worldview that embraces spiritual values that give life its meaning and give humanity a direction for the appropriate application of medical and scientific knowledge. This new paradigm is developed in the next chapter where compassion, rather than science, is the impetus for a new spirit of enterprise to create a better world and improve the human condition.

CHAPTER 9

Compassion:
The New Spirit of Enterprise

The spirit of enterprise, of our individual and collective striving for a better world, has been evident in various modes since time immemorial. These modes express the values and goals of the vision or ethos of the times—tribalism, communism, capitalism, individualism, materialism, industrialism and so forth. Certain virtues and vices are evident in the hopes and sufferings of those living during these epochs of human history and evolution, such as egalitarianism and humility, chauvinism and paternalistic imperialism.

Today, against the dark and gathering shadows of human overpopulation, poverty and famine, and the epidemic of violence against humanity and other sentient life, a new spirit of enterprise that arises from compassion is surfacing. It guides us toward a less violent, less consuming, less harmful, and more fulfilling way of life. The horizon of values and goals is shifting 180 degrees. This radical shift is causing great social, economic and political tensions. It has come like a psychic tidal wave, a tsunami of such luminous power that it could transform the United Nations into a United Environmental Nations. This shift results in a total reorientation in human consciousness: what E. F. Schumaker, the Buddhist economist and author of *Small is Beautiful*, terms a *metanoia*.[1] This metanoia or paradigm shift of sense and sensibility brings bioethics to the forefront of our social, political and corporate agendas, providing the necessary instruments of reason and compassion to help guide and inspire human behavior toward planetary peace and life-enhancing, sustaining, and fulfilling harmony and security.

Bioethical principles—timeless "Victorian" virtues that include obedience to the Golden Rule, humility, honesty, frugality,

139

benevolence, compassion, and reverential respect for all life and for Nature's diversity—can be applied to our relationships with all sentient life, now and forever. They translate into respect for the intrinsic nature (ethos) of all living beings, for their extrinsic value (telos) and for their place in Nature (ecos). Those animal species under our immediate care are entitled to equal and fair consideration. This does not mean equal rights, because humans and other animals have different interests. But it does mean respecting the rights of all animals to humane treatment, and adherence to veterinary bioethical principles: right breeding, right environment, right nutrition, right understanding, and appropriate veterinary care when needed. The *new* spirit of enterprise is directed by a human consciousness that is the antithesis of egocentrism and anthropocentrism because it is ecocentric or Creation-centered—it does not see all of life as a human resource, devoid of intrinsic value.

Creating "manimals" like cows that have human genes so that their milk has more commercial potential and material value, and pigs who have humanized immune systems so that their blood and various organs can be harvested and put into humans, is still the old paradigm at work. Animalized humans and humanized animals: a curious consequence of a spirit of enterprise that uses its powers of dominion for pecuniary and paternalistically self-serving ends, even perhaps to maintain a depraved existence. Many transplant surgeons, some of whom I have debated on public radio and television, see the creation of manimals and the mass production of pig hearts, kidneys, and livers for human recipients as great progress, as one of the incredible fruits of scientific enterprise. It is a great biotechnical feat, indeed. But is it really necessary? Will it not do less to improve the human condition in the long term? Won't treating animals as mere commodities and providers of spare parts and health care products for a sick society further erode our sense of reverence for life?

Against the hype and promises of such scientific enterprise, the deeper human enterprise of bringing the spirituality of compassion into the world can be more clearly articulated. The compassionate spirit of enterprise gives us more holistic vision that sees the reasons why so many people, living in a poisoned environment and eating wrong and harmful foods, continue to live as they do and not change. The victims and perpetrators of great harm and many wrongs are one and the same. When we harm the Earth, we harm ourselves. When we have reverence and compassion for all sentient life, we try our best to harm none. We are then neither victims nor perpetrators.

The enterprise of dominionism brings evil into the world. It seeks to first control and then to conquest and commoditize life, Earth's Creation, Nature's powers, its products and processes. It is ultimately anti-life, compressing biodiversity into monocultures of productive uniformity. In the hallowed name of "progress," it sees industrial productivity, market expansion and increased public consumption as its primary values and goals. Meanwhile, the Earth's metabolic and kalabolic rates are increasing—hence, global warming and the socio-economic and ecological collapse of marine and terrestrial ecosystems, and the annihilation of natural resources.

The spirit of dominionism that brought evil into the world has been around for a long time. It probably first arose when one of our ancestors looked into the fire one windy night and saw a rock melt, and next morning in the cold ashes, found a sharp sliver of iron. That ancestor made the first knife point by sharpening and shaping the sliver against smooth granite and sandstone rocks. Which one of our ancestors used the knife to kill? Which one used it to perform the first surgery, incising an abscess or removing a gangrenous limb? Looking into the electronic fires of their computers, which of our more recent ancestors conceived and developed the life-vaporizing nuclear bomb, and the life-saving radiation therapy: Was it the same mind? It was the same species that decided how to use these new-found powers over life, matter and energy for such different ends. But it was not the same mind.

It is from a very different mind-set, from a different sub-species of Homo sapiens that has metamorphosed from being self-centered to becoming more compassionate and other-centered, that the new compassionate spirit of enterprise arises. From this arising and the associated elevation of both the human spirit and of our awareness and sensitivity, we realize and release the power of compassion. Compassion gives us power over time and space so that others in the future may be well, more secure and freer than we; so that those now less fortunate may be relieved from pain and suffering and from oppression and injustice. Suffering can be prevented through compassionate thought and action.

Like St. Francis of Assisi who empathized so deeply with the suffering of Christ crucified and received the stigmata, so through the spirituality of compassion we receive the stigmata of a suffering animal kingdom and of an Earth crucified by human ignorance and greed. The old paradigm of dominionism—of white male supremacy, and the mythologies of its harmful, mechanistic and reductionistic sciences, public policies and corporate goals—is being superseded. It is being superseded by an awakening of the human spirit to the human condition and to the tragic state of Earth's Creation, so that we no longer live in denial or rationalize further ecocide, or genocide, or the redirection of the universal creative process to serve our own pecuniary ends. This awakened spirit would never strive to remake Earth's Creation, that reflects the *imago dei* to our heightened spiritual senses, into its own image of industrial productivity, commercial capital, and market potential. James Morton, Dean of the Cathedral of St. John the Divine in Harlem, New York, expressed this spiritual vision as seeing ecology "as the body of Christ, through which we of the earth community learn our sacred connections."

Without the bioethical principles of frugality, humility, and benevolence that arise from compassion, the cold spirit of enterprise creates monocultures—agricultural, ecological, social, medical, and commercial. The World Trade Organization is the last creation of this enterprise of materialism and capitalism. If it is to survive and not be the perpetrator and victim of its own nemesis, it must discover the spiritual and ethical aspects of sustainability and the significance and inherent value of others and all beings, of biodiversity and cultural diversity. A global monoculture of industrialism and consumerism, driven by unrestrained capitalism, is the downward spiraling spirit of nemesis, not of enlightened enterprise.

This downward spiral can be reversed, as it must be swiftly, through international cooperation, mutual aid, and a greater mutuality of respect and trust between the rich and poor in every nation, and among different cultures, politics, religions and traditions.[2] What is called for is a United Environmental Nations to harmonize industrial development and market expansion with the needs of the people and the integrity of the biospheric ecosystem, the climatic stability of which is so vital for the maintenance of planetary life and agricultural productivity. What is also needed is more decentralization of power and control, more regional democracy and local self-determination by people educated and informed about how best to avoid making conditions even worse for their children's children than they are now.

Corporate Bioethics

Corporations have developed various operational principles and management directives to keep their products, processes, and services, and marketing claims and strategies within the law. As national and international laws change, reflecting the will, the rights, and the interests of workers and consumers, so corporate ethics have evolved.

As public concerns and influence intensify, corporations and the business community, including banks and stockholders, are being encouraged to increase the scope of corporate ethics to address environmental, social, and ecological concerns. These concerns, if not addressed, could interfere with world trade and result in consumer boycotts, as in Europe's boycotting of genetically engineered food, and lawsuits, as in U.S. consumer group lawsuits against the government for neither properly determining the safety, nor labeling genetically engineered foods and dairy products from cows injected with genetically engineered rBGH.

These examples show that corporations should not operate in an ethical vacuum. By not considering the broad biological—the ecological, public health, social and environmental, and animal welfare—consequences of new products and services at the research and development stage, corporations large and small are going to face increasing public censure, opposition, marketing difficulties, and regulatory obstacles. Bioethics provides the necessary framework for addressing the broad biological consequences of new products and services, that simplistic cost or risk benefit analyses have traditionally ignored or externalized.

A consequentialist analysis raises questions concerning the ecological, cultural (moral), socio-economic, environmental, and animal and human health and welfare impacts of new products and services. The calculus of bioethics includes the following principles and criteria to enable an objective, impartial determination of such impacts:

- In addition to fair market price and veracity of efficacy claims, can the product be easily recalled if defective, or repaired, or recycled?

- Does the product enhance or jeopardize consumer and animal health and welfare?

- Are there demonstrable and quantifiable ecological, environmental, and socio-economic benefits, and what risks and costs are to be accounted for in the calculus of product efficacy and liability?

- How does the product impinge upon human rights (choice, self-determination), transgenerational equity, and even national sovereignty and security?

- Is biological and cultural diversity enhanced or threatened?

- Are rare plant and animal species endangered, and indigenous peoples harmed?

- Are there alternatives that draw less on natural resources, that are environmentally safe, ecologically sound, user or consumer safe, and cause less or no harm to animals both domestic and wild?

The adoption of bioethics by corporations and the business community is, in the final analysis, enlightened corporate self-interest. It is the key to public accountability and respect, if not also to a just society and a sustainable economy.

Use of a numerical scale from 1–10 to judge adherence to these criteria where higher scores would reflect better adherence to each bioethical principle or criterion, could provide an objective tool for corporations to evaluate new products at the research and development stage. Such a tool could also be used for market promotion, like a "Good Housekeeping" seal of approval or the "Green" label of eco-friendly products. Manufacturers would qualify for bronze, silver, or gold labels on products when those products earn an averaged score of 6, 8, or 10, respectively, for a common set of bioethical criteria. This type of labeling has proven extremely successful for various agricultural products and prepared foods marketed in Japan by a national organic food producer's cooperative, although a narrower set of criteria was used.

Certainly some bioethical principles and criteria like social justice and transgenerational equity are challenging, difficult to quantify, but of significant heuristic value. Other bioethical criteria, such as energy costs, biodegradability, ecological impact, and environmental and consumer safety are more easily quantifiable. Establishing an appropriate spectrum of bioethical principles and criteria is an initiative that could help forge a strong alliance between corporations and consumers, provided there is impartial "third party" input (from the Consumers' Union or Union of Concerned Scientists, for example) that has no vested interest in the product in question. Different products and processes raise different bioethical questions, and while some questions may be relevant across the board, others may not be applicable to certain products. But nonapplicability must be demonstrated and not presumed.

The highly competitive nature of the corporate business world is now compounded and confounded, notably by the GATT, the free trade ethos of the GATT's World Trade Organization, by ethical issues concerning national sovereignty, intellectual property rights and human rights, and by product safety and quality concerns now being addressed for global market harmonization by the World Health Organization's *Codex Alimentarius* (an international body that sets food safety and quality standards). Consumers' rights, environmental and animal welfare and biodiversity conservation issues, coupled with the socio-economic, political, and ethical implications of "sustainability" are challenging and formative influences on the competitive spirit of corporate enterprise and a more enlightened capitalism. In a crowded world of finite resources, the call for sustainability, social justice, "eco-justice," and animal rights is the banner under which, I believe, enlightened corporations will flourish in the next millennium. The marketplace will no longer be the primary determinator of what products and services are acceptable. Public acceptability will be based not on convenience, cheapness or personal gain, but on full bioethical evaluation of all so-called externalities, or hidden costs and degrees of consequential harm. For instance in 1998, 280,000 letters to the USDA from a concerned and informed public stopped the government from furthering corporate interests in seeing genetically engineered crops and foods approved under the proposed National Organics Standards for agricultural products. Corporations will lose fortunes if they are not publicly transparent and accountable, and put faith in products that are based on bad science and were developed and marketed in a bioethical vacuum.

We do need to face and fight corruption, from the material to the spiritual, and confront our own and others' selfishness with courage, compassion and commitment, making justice, freedom, peace and the integrity of Creation our collective passion and action. And we must be sincere about it. When we are open, honest, and open hearted, we honor Nature, animals, plants, and our own human natures with loving understanding and respect. We can then accept and nurture the dark, insatiable side of our better natures. This done, we may be entitled to call ourselves Homo sapiens, the wise.

The compassionate spirituality of enterprise may end in self-satisfaction or self-realization, but it begins in service. By the same token, the paths of science and materialism, like the spiritual path, should not end respectively in more power and security. Rather, they should converge, as only paths that rise can converge, to serve all of life in reverence and compassion. The rising paths of the enlightened parent, teacher, scientist, economist, corporate executive and those who have been inspired by the universal light of compassion, all converge. Teilhard de Chardin, in his book *The Future of Man*, called this point of convergence the Omega point.[3] This convergence unifies our senses and sensibilities, our sciences and our ethics, our religions and our desires. Concern for animals and Nature, for eco-justice and social justice, for biodiversity and cultural diversity, become as much a part of the international agenda as human rights, pollution control, and fair trade agreements. The new spirit of enterprise is the spirituality of the new millennium that first demands a full accounting and accountability for the state of the Earth, for the human condition and for the condition of divine Creation and conception. These have been exploited and defiled, causing us and many other fellow beings great harm and much suffering, even extinction.

The end is in sight. It is the end of the old mind-set of human dominion. The new horizon that the spirituality of compassion points us toward—the great enterprise, and challenge, of human evolution—is to become more fully human, humane, and compassionate. Compassion transcends all beliefs, because compassion means action. Compassion is a verb, not a noun. Beliefs can mean action or inaction and cause great harm if there is neither humility nor empathy. All beliefs should be transcended, and are, by compassion. For instance, the narrowed doctrine of ahimsa in India must be transcended when it is believed that to kill an animal for compassionate reasons is to make oneself impure. This selfish perversion of ahimsa is the cause of much animal suffering. This is the tragic situation for millions of India's abandoned "sacred" cows and unwanted, sick and starving male cattle. Such a seemingly compassionate belief, like ahimsa, becomes perverted when it is self-centered and focused on one's own spiritual purity. But when ahimsa is linked with compassion, there is real empathy with other souls and less suffering. It becomes possible to put an end to the death camps where India's unproductive cows and abandoned male calves and bullocks are allowed to starve to death.

The western belief that objective "science" knows best, must also be transcended. When the scientific method is reductionistic and mechanistic it too provides for a limited worldview. When there is empathy, we create an "empathosphere." This "empathosphere," described in my book *The Boundless Circle*, means the end of the self limiting mind-sets of egotism and anthropocentrism and the beginning of a new compassionate spirit of enterprise that captures our imaginations and imbues our lives with purpose, if not hope.[4] This grows as we become more "panempathic," able to have more feeling for and understanding of other beings. An "empathosphere" heals, makes whole, affirms and hallows all beings, and is the source of empathic knowledge. As our ethical and empathic sensibilities are awakened, our mind-set and behavior in relation to other sentient beings are radically changed from exploitation and expropriation to communion and community service. The sense and politics of community are expanded as the communitarian seeks communion, through social justice and eco-justice, with the broader life community of the planet.

While Being is universal, all beings, from the humpback whale in his ocean realm to the spider in her silken web, have their own universe. They have intrinsic value as well as extrinsic biological utility in the co-creative, ecological roles they fulfill. All life forms teach us how to live because they are part of a greater whole—the tree of life—that sustains them, and which they help maintain. This co-evolving, mutually enhancing symbiosis is one of the laws of Nature that we cannot transgress with impunity. Each being is centered in its own reality, which to our senses remains veiled in mystery in the same way that the life universal is an ever unfolding mystery. The politics of compassion translate into nonviolence and an egalitarianism that cuts across all barriers, both social and biological. The spirituality of compassion distills and unifies the essence of all religious traditions into the bioethical ideals of living nonviolently, and with reverential respect for all life.

We can come to integrate all our different needs, wants, beliefs, prejudices and hopes into a coherent framework provided we consider the consequences of our actions, strive to be ethically consistent—neither politically correct nor religiously fundamental—and never act from such a narrow self-interest that we are ignorant, or worse, indifferent, as to how and why we may cause others to suffer. Such ignorance suggests ignorance of one's true self also. The way of compassion means joy and sorrow and a full life that is authentic because it is natural. To be human is to be natural. To be natural is to manifest and experience the sacred, the true nature of the self. As one natural man sees it all, Bill Neidjie, aboriginal elder and spokesman for the Bunitj clan of the Gagudju language group in Australia's Kakadu National Park, says:

> We walk on earth,
> We look after . . .
> like rainbow sitting on top.
> But something underneath,

> under the ground . . .
> we don't know . . .
> you don't know . . .
> What you want to do?
> If you touch . . .
> You might get cyclone, heavy rain, flood,
> Not just here,
> You might kill someone in another country.

He goes on to say:

> "If you feel sore . . .
> headache, sore body,
> that mean somebody killing tree or grass.
> You feel because your body in that tree or earth.
> No body can tell you,
> you got to feel it yourself."[5]

This is the naturally empathic way of the open heart. It is where the spiritual enterprise of rediscovering what it means to be human and redefining what the "good life" means, has its beginning and inspiration. The gift of all this is enthusiasm (*en-theos*—realizing the god within). But first comes humility for all of us humans because we are all part humus, as well as part star dust and light. Though most of us walk upright, we have yet to learn how to walk properly with the open heart of the compassionate and uprighteous. Vietnamese Monk Thich Nhat Hanh says, "The miracle is not to walk on water, but to walk on the Earth."

POSTSCRIPT

Animal Rights and Human Liberation

Animals can't answer for themselves, so who is to speak for them? Those who speak for them, using legalistic rights language, will never get very far because such language is locked within the mind-set of Greek (Aristotelian) rationalism and Cartesian dualism.[6] Because these spokespersons believe humans are superior, their reason informs them that animal rights are subordinate to human rights and interests. Furthermore, the question of animal rights is judged from a point of reference that is derived from Roman law and jurisprudence. According to this original jurisprudence, the violation of animals' rights is acceptable if it benefits the empire or society in any way.

This Graeco-Roman rationalism found fertile soil in Judaism and Christianity arguably because of their anthropomorphic conception of divinity and consequential anthropocentric and chauvinistic worldview—and because the leaders of these faith traditions were human and therefore fallible and corruptible. It was never beneath most priests to slaughter and eat the fatted calf and invest in livestock, or as a religious order to invest in real estate or in war, and today to speculate in the stock market. Consequently, the voice of these monotheistic religious traditions has been almost silent in speaking out on behalf of animal rights. This was not, I believe, to avoid the Aristotelian trap that rationalizes human superiority over all creatures and Creation, but in order actually to preserve the status quo of animal utility and human superiority.

In order for animal rights to prevail and be accepted by all, it must come from an ethical basis that is spiritual rather than legalistic or moralistic. It must come from the heart of compassion and from reason that proclaims that we and all creatures are related, part of the same divine Creation and conception. Otherwise the necessary spiritual transformation of humanity will not be accomplished by those who seek to defend animals and protect them from our inhumanity. From this perspective, the spirituality of animal rights is a challenge to both organized religion and western civilization. It is a call for a new religion, ethics and social order.

The discovery of our humanity is a spiritual process, not one of law and order or rationalism and scientism. The ultimate task of all who still have empathy for animals and are concerned about the fate of Earth's Creation is the recovery of our humanity, our compassion, humility, empathy, natural wisdom and reverential respect for all Creation.

"Feral" Vision

Our perception of the world—how we see other life forms, Nature's creations and "resources"—is influenced by how we value and feel for and about them. When we rid ourselves of all cultural programming and selfish wants and expectations, our perception changes. When we are loved by an animal, as when we love an animal, or a tree, or some natural place, our perception changes. Too often that perceptual change is fleeting and we quickly return to our habitual ways of perceiving, being, and relating. We may or may not be aware of the splitting or separation that occurs when we shift from one way of seeing to another. The irony and ethical inconsistency may escape us of situations like the woman who loves her cats but wears a wild cat fur coat, and the biomedical research scientist who plays with his golden retriever before he goes to the laboratory to experiment on dogs. The ultimate irony for ethical vegetarians is that people say they love animals but eat them nonetheless.

Many rationalists, often with a science background, believe that life is not a process and expression of anything sacred or divine, but is a consequence of pure chance. Less thoughtful materialists or avowed atheists must deny or sublimate any feeling for the sacred, the *mysterium tremendum* that arises naturally from childhood's sense of awe and wonder. Parents, teachers, peers, and culture too often combine and conspire to crush this childhood gift of primordial feral perception. Walt Whitman described so eloquently this way of seeing and being as a child who became everything he beheld.

It is on the basis of this primordial feral vision that our sense of kinship with all life develops, along with the sense of self and the capacity to empathize. From empathy such human attributes as compassion and ethical sensibility arise naturally. When the development of feral vision is inhibited, the development of the whole person is impaired. Morality and law and order must be imposed from without when there is no internal, spontaneously arising empathic sensitivity and ethical sensibility. In such a society as we have today, feral vision leads to spiritual anarchy and the call to liberate all creatures and Creation from the inhumane and selfish dominion of those who see the world simply as a resource, as matter without spirit, as theirs by divine decree. The whole of Creation is not a means to human ends. Every being is an end unto itself. As Thomas Berry sees it, the universe is not a collection of objects but a communion of subjects.

World peace, justice and the integrity and future of Creation will never be secured with a purely anthropocentric perspective that sees Nature as a resource and that has no sense of kinship with animals. With feral vision the spiritual anarchist sees peace, justice, and the integrity and future of Creation from an anthropocosmic perspective. This more holistic mind set entails a paradigm shift from the conventional, human-centered worldview. This "shift" is difficult to achieve when the primordial, feral perception of childhood has not been nurtured and has instead been inhibited and allowed to atrophy.

The recovery of feral vision is linked ultimately with the recovery of our humanity. Sometimes this vision is restored by great personal tragedy, by near fatal illness, a near death experience, or by the devotion of an animal. There are religious and spiritual traditions, some old and some new like the vision quest, that can help us regain this way of seeing and of being. Those who lack such vision cannot know what they are not seeing and feeling until they have gone through the paradigm shift. Only then can they look back on who and what they were before their vision was restored.

How is this shift in consciousness and perception achieved? As prevention is the best medicine, it is best to nurture the primordial, feral vision in childhood via humane and environmental education and spiritual guidance, to facilitate the development of empathy and ethical sensibility. But when we have an entire industrial civilization based on the values of materialism and consumerism, and economic and political forces that maintain the status quo and collective mind set, then the possibility of a metanoia, a radical change in consciousness and perception, seems infinitely remote. It will be until those with feral vision, the spiritual anarchists of this age, begin to challenge the status quo and show, for instance, that the economy is as unsustainable as it is socially unjust. In the process these spiritual anarchists will provide consumers with an alternative vision that empowers them to see that the corruption of politicians is merely a reflection of the corruption of the spirit of contemporary society. Like the child in Hans Christian Andersen's story who proclaimed that the emperor was not wearing any clothes, so a populace, with feral vision, can see the sham and artifice of those who seek power and profit at the expense of their own kind, of animal kind, and of Earth's Creation. In Nature's image and living presence, we have our being and becoming. In God's image, Nature is, and every sentient being and natural form, process and phenomenon, from the atoms of matter to the genes of life, a reflection of divinity.

The spiritual worldview is normative for those who have not lost their feral vision. Little wonder such people feel alienated from the consensus reality their peers and elders embrace—the reality that condones the commoditization of all life and sees nothing sacred or of significance beyond the values and aspirations of a purely materialistic, consumption-oriented existence. When those with feral vision make their views and concerns known, but rarely understood, they are too often ostracized and ridiculed as being unrealistic, idealistic, and anti-establishment. In past ages they were variously shunned, excommunicated, persecuted, and even burned at the stake.

It is inconceivable for those who defend the status quo of inhumanity to accept that this status quo is the very cause of all our worldly tribulations and sufferings. But on the grounds of reason and compassion, any rational being would take every step to change it. The status quo speaks of rights but denies them, speaks of justice but is unjust, speaks of morality but has a self-centered life-ethic. The status quo has no values higher than materialism, no horizon beyond the acquisition of wealth and power. This "system" cannot change itself. We and our forefathers created it. So we must change it. As more people of all ages "go wild" and reclaim their humanity, their authenticity, and significance in life, with feral vision and the spiritual anarchy that arises therefrom, this gentle revolution and the evolution of our species will begin.

It is beginning now, this change in perception, in our consciousness and how we relate to and value other sentient beings and the living Earth. I call this the beginning of the age of ethics, the 'Ethicozoic' age. Brian Swimme and Thomas Berry capture this self-transformative process through their own feral vision of an emerging Ethicozoic age that they call the Ecozoic era (the awakening of ecological awareness). They write: "That the universe is a communion of subjects rather than a collection of objects is the central commitment of the Ecozoic. . . . We need an inter-species economy and an inter-species wellbeing, an inter-species education, an inter-species governance, and an inter-species religious mode, inter-species ethical norms." The animals, the trees, the Earth and all living beings do speak for themselves when we listen and pay attention. And it is better not to ask "Are you for or against animal rights?" but rather "Are you for or against peace, justice, compassion, and the integrity and future of Creation?"

Our species is now at a biological and evolutionary crossroads. As Teilhard de Chardin observed some thirty years ago in his book *The Future of Man*, we have reached the point in our biological evolution where we have one final choice to make. That is between suicide and adoration. When the world economy, and the politics, policies, and priorities of nation-states cease to operate in an ethical vacuum, totally devoid of reverential respect for all life, then we will be more secure. Only that which is sacred is secure. And from the hindsight of feral vision we see that reverential respect for all life is enlightened self-interest. It is the highest value to live by and the only one worth dying for. In our recognition and acceptance of animal rights, not as some legal or moral principle, but as a spiritual and ethical imperative, we will realize our own liberation.

Animals on the Global Agenda of Human Concern and Responsibility

The rights of Nature, humans, and animals are interconnected and interdependent. Animals fulfill many important roles in society as wild animals do in natural ecosystems. Their health and welfare are inseparable from the health and well-being of society and of Nature. They are thus worthy of our respect and humane treatment. Animal cruelty and suffering should be prevented in all countries for the moral and ethical integrity and future of society. And animals' place in Nature should be respected for the ecological integrity and future of Nature. In the next chapter we will examine the validity of the claim that caring for creatures and Creation is the key to a humane and sustainable economy and global community.

CHAPTER 10

The Life Ethic of a New Covenant for a
Sustainable Economy and Future

Recent developments in biotechnology and in other technologies
have been touted as panaceas for failing industrial economies, and
as sources for greater agricultural productivity and cures for the diseases of civilization. A closer look at biotechnology in particular gives
little grounds for such optimism, however, especially since there are
no signs that multinational corporations and governments alike
sense any urgent need to change their industrial economic paradigm
or worldview.[1] A new paradigm or covenant is sorely needed since a
"business as usual attitude," and unqualified emphasis upon such
subjective criteria as industrial growth, productivity and profitability, are incompatible with the good of society and the future well-
being of the life community of this planet.[2]

The widespread use of public funds for government and university research and development has done more to benefit the private
corporate sector than to further the public good. No better example
comes from the U.S. than the Land Grant University system that
has become a branch of the petrochemical, food, and pharmaceutical industrial complex, through endowments, public tax moneys,
and mission-oriented research funded by the private sector. At the
same time the economies of rural communities, which the universi-
ties are mandated by Congress to aid and develop, have collapsed
around them.

Likewise, in part because of intense lobbying and the providing
of moneys (and even weapons abroad) to politicians to secure reelec-
tion, multinational corporations have gained a deep and pervasive
influence over governments worldwide. As a consequence, it seems
that even in the most benevolent guise, the best intentions of good
men go awry.

153

It is remarkable and disturbing that many, many people, including those in positions of responsible authority and influence, are like members of a dysfunctional family, in a state of denial over the dysfunctional and rapidly deteriorating condition of industrial civilization.

Indeed, the denial of the dysfunctional state and harmful consequences of the petrochemical, food and pharmaceutical industrial complex by agribusiness, academia, the medial establishment, and other "family" members with a vested interest in preserving the status quo is a conspiracy of silence that history will record as one of the greatest evils of the industrial age.

This situation has arisen in part because of a naive enchantment with technology, and seriously flawed cost accounting. Similarly, aid and development programs, funded by the World Bank, International Monetary Fund, and other publicly subsidized institutions, and backed by the "scientific" authority of academia, primarily to facilitate the colonial expansion of corporate interests, have caused great harm to the delicate socio-economic and socio-ecological integrity of most Third World countries. The billions of dollars in development loans have caused the Third World to incur an enormous financial burden, which has led many countries to export nonrenewable resources and sacrifice their natural heritages simply to pay the interest on these loans.

Economist Herman E. Daly points out that "when economic growth exceeds the optimal scale [of sustainability] we experience generalized pervasive externalities, such as the greenhouse effect and acid rain, which are not correctable by internalization of localized external costs into a specific price."[3] Daly also voices his concern that growth, fostered by world free trade, may impose unacceptable hidden costs by eroding the economic basis of national communities and limiting their ability to set their own social and environmental standards. He writes, "Once community is devalued in the name of free trade and global integration, there will be a generalized competing away of all community standards that raise costs of production . . . Free trade, as a way of erasing the effect of national boundaries is simultaneously an invitation to the tragedy of the commons."

This "tragedy of the commons," a term coined by Garett Hardin, refers to the inevitable destruction of any ecosystem that is not used sustainably. The United Nations report, *Our Common Future*, defines sustainable development as "meeting the needs and aspirations of present generations without compromising the ability of future generations to meet their needs. It requires political reform, access to knowledge and resources, and a more just and equitable distribution of wealth within and between nations."[4]

The overemphasis of industrialism on economic growth and "progress" has caused great harm to sustainable communities worldwide. Theologian John B. Cobb, Jr. has termed this ideal of economic growth *economism*. Its end result in agriculture is factory farming. Economism equates economic growth with the ultimate good of society—more services and consumables, and more productivity for less labor. Economism's hallmark is efficiency.

Economism is not a benevolent worldview, especially when contrasted to what Cobb calls *planetism*. Planetism embraces such bioethical principles as sustainability, social justice, concern for animal welfare, and environmental protection. To be adopted, planetism must begin locally and bioregionally. It must follow the image not of one corporate global village but of a globe of many villages, composed of self-reliant and sustainable communities whose economies do not impoverish others, do not harm the environment or other animals, and which strive to enhance the well-being of other communities, both human and nonhuman.

Economism commoditizes animals (as goods or products), dehumanizes the human communities (as a labor force and consumer market) and makes the natural world less sacred by judging it as a mere resource. In contrast, the economy of planetism regards the earth as sacred, affirms the dignity and worth of the human community, and extends compassion and respect to all creatures. It is upon these principles that a humane, sustainable, socially just, and viable global economy can be built.

Edward Goldsmith in his book *The Way* observes that "the progressive degradation of the biosphere which we are witnessing today cannot be attributed to technical deficiencies in the implementation of our socio-economic policies. It is the policies themselves that *by their very nature* are causing destruction."[5] These policies are based upon an attitude toward life and toward Nature's resources that is incompatible with the well-being of both the human and nonhuman life communities of planet Earth. Industrial growth and economic development are sanctified as "progress" for the good of society, yet they have become the undoing of both society and the natural world, upon which the human community is both dependent and beholden. The harmful consequences of these "good works" of well intended people have the same bathos as those of people who work for, and invest in, industries and business activities that ultimately harm them, and that are antithetical to their basic needs and entitlement to a socially just society, a healthful environment, and a secure and sustainable future for their children. This schizoid condition is rationalized and denied, intensifying the stress and pathology of the human condition, and of the biosphere itself.

Ecological Wisdom and Policy of Compassion

In many instances, we lack sufficient science-based knowledge to determine public policy over issues like global warming and the linkage between loss of biodiversity, decline in climate regulation, groundwater replenishment, and biodegradation of anthropogenic pollutants. Hence, the need for a broad-based life ethic that provides an ethos, or guiding value, to society. This ethos is not based upon or bound by scientific, legal, or economic principles and rationales. Rather, these considerations are secondary and should arise from the guiding ethos of bioethics. Otherwise, we will continue to put the cart before the horse. In sum, society must replace legislation with ethics as the primary point of reference and means of ensuring the greater good of the life community as a whole.

It is surely a flaw of dualistic thinking to believe that when we harm the environment we can do so without causing harm to ourselves. The new field of environmental medicine reveals clearly how our physical well-being and the well-being of future generations is very much a function of how we perceive and treat the environment. The declining quality and safety of our food, air, and water are reflections of our collectively harmful way of thinking and behaving.

Another flaw of dualistic thinking is our lack of appreciation that when we harm others, including animals, we harm ourselves. Acts of violence and public acceptance of cruelty toward animals leads to a callous indifference and a withdrawal of empathy and compassion. And we harm ourselves in the process. The psychophysiological, cognitive, and affective consequences of cruelty toward other sentient beings, and indifference toward their suffering, are surely damaging to the human spirit. As compassion is a boundless ethic, so violence knows no bounds when there is no empathy, no felt connection with other sentient beings to enable us to live gently and mindful of others. Thus an inhumane society is as spiritually sick and physically violent as a society that sickens itself through environmental destruction and contamination.

A new relationship, or covenant, is needed in our dealings with the land and the life around us. The bioethical and spiritual basis of this relationship entails a reorientation of our attitudes and values from those of objectification and exploitative commoditization to instead subjective communion and creative symbiosis. The maintenance and functional integrity of the planetary ecosystem, or biosphere, of its oceans and forests, as well as of our own communities, urban and rural, are disintegrating and becoming increasingly dysfunctional because of the prevailing worldview and the way we structure reality.

It is by way of empathy and compassion that a new covenant with Creation is to be made, and our instrumental scientific knowledge best utilized. Empathy and compassion will reawaken our intuitive wisdom to heal the earth and ourselves in the process. We then shift from a competitive, adversarial, dominator mode of being to one that is cooperative, nurturing, and creative in *all* our relations.

This fundamental shift is an essential developmental and evolutionary step for the human species. There is no alternative, unless we choose to be a terminal, exterminator species. Care and concern precede ethical choice and action. Thus, compassion and empathy are prerequisites to ethical behavior.

Compassion and empathy broaden the scope of our concerns such that we do not lose sight of the future that we are creating when we focus on finding solutions for our myriad problems. Even while we acknowledge the capacities of good and evil in our own chimeric and contradictory nature, and which the struggle for survival and co-evolved interdependent symbioses in all of Nature reflect, we humans have the power of choice: that of good over evil, nonviolence over violence, and of giving more to life than we take. And so we must if the Earth is to sustain our burgeoning numbers.

Charles Darwin, father of the theory of biological evolution, was one of the first scientists to identify that the law of Nature—based upon completion and the survival of the fittest—is integral to the process of natural selection. Even so, he was quite adamant in his opinion that humans are not superior to the rest of the animal kingdom, from whom we differ, he maintained, only in degree and not in kind.

With this now human-endangered living being, planet Earth, we must make a new covenant to establish a mutually enhancing relationship. This relationship cannot be based upon the continuance of the prevailing attitude that has, as Donald Worster documents, reduced plants and animals to insensate matter, devoid of internal purpose or intelligence. Mechanistic science, he contends, has removed the remaining barriers to unrestrained economic exploitation.[6] While for industrial society the order of Nature is one thing and the social order is another, to the Australian aborigine they are part of a single order.[7]

Holarchism—A New World Order

Worthy of consideration is the nonviolent anarchism of Peter Kropotkin, in his book *Mutual Aid*. He attacked Thomas Huxley and other Social Darwinists, insisting that cooperation and mutual aid were the norms in both natural and social worlds, and not primarily a competitive survival of the fittest. He wrote, "Mutual aid is as much a law of life as mutual struggle, but . . . as a factor of evolution, it most probably has a greater importance in as much as it favors the development of such habits and characters as ensure the maintenance and further development of the species, together with the greatest amount of welfare and enjoyment of life for the individual, with the least waste of energy."[8] The erroneous belief, held as scientific fact, that Nature is "red in tooth and claw" and that only the fittest survive, gave credence to competitive individualism in industry and commerce as something natural rather than pathological.

More recent scientific studies reveal that Nature does not function in this way. Paradoxically, competition between species over millions of years almost invariably results in the evolution of a cooperative harmony. The prey-predator relationship of deer and wolves, which existed long before human beings first evolved, is competitive from an individualistic viewpoint. But from an ecological, rather than a human-centered perspective, it is a co-evolved, co-creative, and cooperative relationship. The wolves, fewer in number than the deer, regulate the deer population; without wolves, the deer population would explode beyond the carrying capacity of the environment, and subsequently die out. As the wolves keep the deer herds healthy (like the gatherer-hunters and agrarians who learned to live within the carrying capacity of the environment), so the deer sustain wolves. This relationship is ultimately one of cooperation. Each serves the other.

The competitive worldview of Social Darwinists meant hierarchy, conflict, and violence, and played a central role in the politics and philosophy of the industrial revolution. Enlightenment beliefs of the seventeenth and eighteenth centuries remain, incorporated into liberal and social democracy, especially in its equating a competitive market economy and industrial growth with social progress. And

central to this is the persisting Enlightenment belief in man's superiority within the natural order, and in a secular, empirical sense of
human progress and perfectibility where state rule is considered the
proper and rational instrument of progress.

Because of the negative associations of anarchy with fascism
and bolshevism, a more temperate, contemporary alternative for
anarchy might be termed holarchy. Holarchism is a holistic and non-
dualistic worldview based upon the bioethical ideals of a sustainable, humane, and socially just global community. The political
expression of bioethics is transpecies democracy. It is closer to the
nonviolent anarchism of a self-actualizing and self-organizing community than it is to an authoritarian, hierarchical oligarchy, as exemplified by state and corporate capitalism of socialist and
quasi-democratic governmental technocracies.

The thesis of anarchism is that no government of the people is
necessary because, like all wild creatures, flowers and grasses of the
prairie, they are self-governing with co-evolved, mutually enhancing
relational contracts within a cooperative holarchy. What may appear
to be competition between species (as between predators and prey,
herbivores and plants, parasites and hosts) is actually a beneficial,
highly evolved dynamic state of reciprocal restraint and maintenance that maximizes biological diversity and the stability of the biotic community.

Humans are not the sole members of the ecological or life community of the planet. Human and nonhuman life forms are part of
the same Creation and Earth community. The democratic recognition and consideration of the rights and interests of both human and
nonhuman life forms is the hallmark of a truly civilized, compassionate society. The concept of transpecies democracy is in the hol-
archist's political expression of bioethics.

We must now go forward to establish a sustainable, ecologically
sound, and socially just global economy that is based on an Earth- or
Creation-centered, rather than human-centered, ethical and spiritual
sensibility. We must acknowledge the pathology of separating, or objectifying, and commoditizing life to serve exclusively human ends,
and must heal our relationship with the rest of Earth's Creation, for
the ultimate well-being of ourselves and of all generations to come.
This healing entails not simply abandoning all forms of animal and
Nature exploitation, or in developing less harmful technologies and
better legal restraints and alternative economic incentives. It also entails coupling a sympathic reverence for all life with the ethic of
kindly use, as theologian Jay McDaniel has eloquently proposed.[9]

Kindly use is founded upon respect for the intrinsic or inherent value of animals and all living entities—including functionally whole ecosystems—and recognition of the fact that they are ends in themselves and should not be treated simply as a means to satisfy purely human ends. This change in relationship and healing process also involves what Thomas Berry calls a communion of subjects. This is the antithesis of a commerce of objects where the commoditization of life as a means to our own ends leads to destructive exploitation rather than creative participation.

All living beings have extrinsic, instrumental value in terms of their contribution to the functional integrity and wholeness of ecosystems. As one life is consumed by another, so life is sustained, held in balance and diversity preserved. In other words, life is in the service of life. The ultimate telos or final purpose of every life form is its contribution to the functional integrity of the planetary life support system. It is too simplistic and hierarchical, to say that one life form, like a deer, is simply a means to another's end, like a wolf. The element of reciprocity between species actually transcends the apparent elements of exploitation and competition to serve the greater good—the integrity, vitality, health, and harmony of the ecosystem.

It is this principle, this "lesson from Nature," that must be incorporated into the ethic of kindly, and sustainable/renewable use of other living beings and life processes. By this, we humans will become fully aware that the greater good is not the good of society, or the wealth of the nations, separate from the preservation and restoration of the natural world. The antithesis of good is evil. In the next chapter, a new perspective is given to our understanding of the nature of good and evil, in the belief that without understanding, evil will continue to flourish while good people do nothing.

CHAPTER 11

Good and Evil:
A New Perspective

Predator species, like the lion and the wolf, are the interlocuters between chaos and order, playing a vital role in keeping the herbivore species in balance with the carrying capacity of the vegetation. Many plant species have coevolved with herbivorous mammals, insects, and birds; predators have coevolved with herbivores; and scavengers, carrion eaters, and decomposers of organic material have also become integral parts of the ecosystem. This coevolution forms a holarchy—a holarchy that is amoral, meaning that it lies beyond right and wrong, good and evil. What is right and good for the wolf is also good for the deer. The death and brief suffering of the deer is not evil or wrong, since the balance of Nature and carrying capacity of the whole ecosystem is maintained. Too many herbivores would mean annihilation of vegetation and a consequent slow death for these animals from starvation and disease.

From this naturalistic perspective, what is wrong and evil is that which upsets the equilibrium and integrity of the ecosystem's holarchy, and what is right and good is that which allows all life forms within the ecosystem to achieve their fulfillment or ultimate purpose. This self-actualization or *entelechy* of plants and animals ensures balance, harmony, and renewal. The only evil in the world emanates from us, and evil is a matter of our choices and awareness.

Animals and all living beings manifest an intrinsic goodness in contributing to the maintenance of the holarchy and life community via entelechy, where each living being contributes to the good of the whole. The law of entelechy and the intrinsic nature, or ethos, and the telos of animals and plants prevent entropy or the chaotic disintegration from taking over the entire holarchy. But at the micro-level of decay and disintegration in the regenerative life cycles of every natural ecosystem, entropy is a vital regulatory process.

From the perspective of biosophy, our human-oriented morality should not regard all suffering and death as evil or immoral. Natural, pervasive suffering of sentient life and death are neither good nor evil, since they are integral to the life experience and process. But death and suffering that is caused by human, rather than natural means—for example war, pestilence, or famine triggered especially by overpopulation and overconsumption—is evil. The death and destruction of animal and plant communities in the process of expanding our own community and our crops and flocks, orchards, and herds is evil when the natural rate of entropy is increased. The entelechy of other species—predators, competitors, "weeds"—is denied when they are exterminated, while the entelechy of "useful" species is directed to satisfy primarily human ends.

Admittedly, one may argue that since the deer is a means to a wolf's end, then there is nothing wrong with domesticating animals and plants to satisfy purely human ends. This is morally acceptable provided no evil results, evil as that which causes suffering to sentient beings and harm to the holarchy of the ecosystem by disrupting the natural rate of entropy, and diminishing biodiversity, equilibrium, and regenerative potential.

Some would argue that since suffering is suffering, there is no real difference between *natural evil*—the inevitable suffering of sentient beings from extreme cold, heat, drought, periodic population explosions (often caused by human activities), disease, injury, or starvation—and *human evil* that arises as a consequence of human need and want. But the difference is clearly evident in the primary cause or source. When we are that cause or source, we surely have the power to choose to cause less evil by finding less harmful ways to meet our basic needs and by foregoing various wants that violate the doctrine of ahimsa (see Chapter 13).

When we violate such natural laws and processes as diversity, self-renewal, and interdependence, we also violate and disrupt balance and harmony. New laws must then be imposed from without, like those dealing with industrial pollution, habitat, and endangered species protection, and external correctives must be put in place, like fertilizers, pesticides, and pest controls. But until the frame of reference for good and evil is enlarged so that every life form is equally included within the scope of our moral consideration and concern, and until we are able to define, from a biosophical perspective, what is good and evil, we will never attain full moral maturity. As a result society and the natural world will become increasingly dysfunctional.

A *"moral majority"* continues to interpret those things in Nature that we don't like in society—such as suffering and death—as evil, and variously blames God, or Satan, sees the natural world as "fallen," and justifies its own evil acts against each other and other life forms as natural and necessary. Another rising *"immoral majority"* objectifies the world and sees other life forms as objects and not as subjects. Some believe their God made the world for them to conquer and rule, while atheistic materialists simply see other life forms as a means and not as ends in themselves. Many are agnostics or atheists because they see God as unloving for making a world that is never free from suffering and natural disaster.

A refined definition of good and evil based upon biosophy will do much to help us attain moral maturity, to heal ourselves and the natural world. But we must first acknowledge that the evil in the world emanates from us and that it is all a matter of choice and awareness. In the process of transforming the natural world to fit our own image of perfected utility, humankind has brought more evil than good into the world by destroying other life forms and their habitats. As a protean, transformative species, we have the capacity to reverse this process and bring more good than evil into the world and to alleviate, as best we can, the natural suffering that afflicts all sentient life, which has yet to be liberated from all forms of human-caused harm and suffering. In order to accomplish this, we need to restore what is left of the natural world, rather than continue to transform it into a bioindustrialized wasteland. We must also restore our own sense of humanity by transforming ourselves into more compassionate and creative participants in the life community of the planet. We do this through establishing a covenant of mutual enhancement.

Changing Human Nature

Except for what we have done to Nature, there is nothing wrong with Nature. Similarly, except for what we have done to our own nature, there is nothing wrong with human nature. What we have done is to create a mind-set or worldview that is harmful to both Nature and to ourselves. And so long as we continue to operate from this state of mind, we will never enjoy the peace and security, or the freedom and fulfillment our hearts desire. Since the human heart is so closely bound with our ancient instincts, with empathy, compassion, and with our moral sense and ethical sensibility, it is the best agent we have to help us change our minds, to make the paradigm shift so that our worldview is as world-sustaining as it is self-fulfilling.

Distrust of Nature is evident in the prevailing negativity toward human nature that has reached the point where the specter of eugenics is again upon us. Genetic determinism and biofascism go hand in hand. For example, the genetic enhancement via genetic engineering of plants, animals, and humans is accepted by many. Genes responsible for human aggression, sexual behavior, and other traits are being identified and sought in order to improve humanity. Also sought are a new generation of pharmaceuticals to control human behavior and emotions. This genetic approach totally discounts the role of "Mother Culture"—the source of nurturance and the social environments which are primarily responsible for human well-being and for violence, despair, and insanity. Such massive denial on the part of conventional medicine could be defended on the grounds of profitability and feasibility: no drug or biotechnology company would profit from changing the social environment, which many people feel is impossible anyway. The approach that is predicated by a negative attitude toward both Nature and human nature will ensure that many of the next generation will live in the ruins of industrial society and in a ruined natural world.

It is ironic that we have been unwilling to adapt to Nature and practice self-restraint, or self-governance, to restrain our competitive and destructive expansion. Instead, ignoring the bioethical principles and virtues derivative of a deep understanding of natural law, we have changed Nature to accommodate our own needs and wants. Now is the time to change society to accommodate and restore both Nature and human nature. The pathological condition of both is a deepening problem, which neither biotechnology nor any other technology can ever correct.

We cannot continue to ignore the gravity and severity of the human condition and blindly try to create a Wonderworld out of new technologies that violate the ecological tenets of natural law. We cannot continue to apply economic and political remedies and reforms that are all "fixes" that do not change but rather perpetuate an untenable worldview and a worsening world situation. Rather the answer lies between us and Nature, not simply between human nature and human nurture. When our empathy and compassion extend to all of Nature, the solutions to our many problems will be forthcoming because our world and worldview will be radically changed for the better. In other words, when we are not in our right minds, the right solutions will always evade us.

Nature's Nemesis of Human "Progress"

Nature is often mistakenly blamed for much human suffering caused by floods, droughts, famine, and pestilence, while in fact these calamities were brought on by various human activities. To understand the meaning of *nemesis*—inevitable retributive justice—we need only to see how the industrialist and materialist ideals of human progress almost invariably and so often with the best intentions backfire, causing greater harm than was ever anticipated. The harmful consequences of the ignorant abuse and arrogant misuse of antibiotics and pesticides in conventional medicine and agriculture are good examples of how Nature's nemesis is invoked. Pesticides harm our health, contributing to cancer and genetic disorders, and the misuse of antibiotics has led to antibiotic-resistant bacteria, with devastating consequences. When we harm Nature, we harm ourselves. One of Ivan Illich's books entitled *Medical Nemesis*, and my own book *Agricide*, show how a rising global corporate oligopoly—the petrochemical, food and pharmaceutical industrial complex—pushes and weaves its products, processes, and services into social and market economies of developing countries, which were formerly more equitable and sustainable.[1] Developed, industrialized nations—biotechnocracies, governing expanding military-industrial complexes—foist their way of life on the rest of the world. They hope to legally enforce their way of life via the General Agreement on Tariffs and Trade (GATT) and the World Trade Organization (WTO), the World Bank having been one of the most faithful missionaries for both these global associations.

Capital input enabling industrialization provides some employment for former pastoral, agricultural, and coastal fishing communities. These once viable, and for generations extremely sustainable, cultures are being obliterated by the industrial technocracies, notably via factory fishing, factory forestry, and factory farming. These are responsible for the collapse of ocean fish stocks, deforestation and global climatic changes, serious pollution from agrichemical fertilizers and pesticides, and global loss of top soil quantity and quality.

Human overpopulation results in ecological devastation when the numbers of people exceed the carrying capacity of the land.

There is also increasing conflict over land rights and access to natural resources, exacerbated by ethnic, religious and tribal differences, leading to increasing numbers of people becoming landless and homeless. The world's looming humanitarian disasters involving more than two dozen nations and a conservative estimate of 40 million people at risk of malnutrition and death mean that relief efforts, according to a classified U.S. Intelligence Report, will not be able to keep up with increasing global conflicts.[2]

The more fortunate of these disenfranchised people are obliged to seek employment in any industry when the opportunity arises, regardless of risks and working conditions. (Many multinational corporations have capitalized on this cheap labor source.) The alternatives of dire poverty, despair, and hunger are driving people to become dependent on the very system that has put them in this choiceless situation. Low-cost employment is encouraged since too much automation means overcapitalization and fewer jobs and thus fewer people able to afford technical and medical services, and the latest consumer goods like costly and wasteful sirloin steaks and fast cars.

The upper class expects these things and ensures them for itself. The more scarce and costly they become, the more the ranks of the middle class swell. As resources become even more scarce and the full costs of production and overconsumption are accounted for, like environmental pollution and depletion, cancer and atherosclerosis, this middle class feels the ever increasing burdens of inflation, rising health costs, failing industries, loss of jobs, and myriad other stresses and strains of a dysfunctional society in the grip of Nature's nemesis. The lower class is in some ways better off than the middle class, since for several generations in the poorer, more ravaged parts of the world, its members have gained considerable experience in learning how to survive in a dysfunctional, post-industrial society, and practice the "3R's" of sustainable subsistence—repair, reuse, and recycle.

As our numbers continue to expand, our expectations of how well the Earth can meet all our needs and wants must be lowered. Tragically, this lowering will not happen fast enough if China, as it industrializes and enters the world market, raises its expectations and buys up America's grain so its people can eat as much meat as westerners do today, and manufactures and imports more refrigerators, air conditioners and faster cars. A wise Indian government would move public opinion to ban McDonald's and Kentucky Fried Chicken from its shores. A more enlightened U.S. government and urban populace would act in unison to stem the demise of its own rural communities. Nearly half a million sustainable family farms have been lost in the U.S. over the past decade. By preventing the spread of factory farms and the monopolistic control of agriculture by a handful of multinational corporations, such agricide would be averted.

A sensible World Trade Organization and its food safety advisory group, the Codex Alimentarius, would put environmental protection, the phasing out of harmful pesticides, and international criteria for certifying organic farming methods and produce, at the top of their agenda. Governments, industries and trade organizations will never develop the bioethical policies and principles of a humane and sustainable world community and market economy until we all lower our expectations to a level that the Earth can sustain. This is the only way to achieve world peace and to avoid Nature's nemesis. It entails redefining what it means to be human and what we mean by progress. From a spiritual or ethical and ecological perspective, this surely means restoring and protecting the natural world and respecting the rights and interests of indigenous peoples, since the more we exploit and harm, the more we will suffer Nature's nemesis. There are grounds for hope since the human species is devoid of neither reason nor compassion. But first we must in all humility accept the wisdom of the doctrines of *ahimsa* and *karma* and acknowledge that no good can come from evil means. Nature will not serve us if we do not serve Nature.

On our present evolutionary path, the extinction of the natural world, and of the human species, in spirit if not also in body, seems inevitable. Our collective evolution parallels the development of the individual, who in childhood gains self-esteem through parental love, and self-control and respect for all living beings through the example of elders. This developmental process is the path of spiritual growth leading to an understanding of the nature of reality, the nature of Creation, and the nature and tangible reality of the Creator. This human path is neither superior nor inferior to that of other life forms. Rather, it leads us to a more encompassing conscious realization of the interconnectedness, interdependence, and mutuality of origin of all life. This path leads us toward reverential respect and a hallowing way of life as we become more compassionate and realize our healing powers and the higher empathic powers of love, wisdom, and humility. At this stage of self-realization and community actualization, we live in accordance with the Golden Rule. We do so, though, not out of obedience to some externally imposed laws or moral code, but because we have become panempathetic, emotionally connected with all sentient beings. This is the affective dimension of a panentheistic worldview.

From this evolutionary and psychohistorical perspective, it is evident that the human species is at an impasse, caught somewhere between the infantile and adolescent stages of development—and Nature is doing her very best to help us mature. And we will, provided we teach the next generation by our own example that self-control is necessary in order to avoid causing harm to other living beings and evoking Nature's nemesis. While self-control may be anathema to an overconsumptive consumer society, and the notion of reverential respect for all life may be meaningless to an immature, egotistical materialist, boundaries must be set. Obedience to Nature's laws and conventions to protect animals and the environment must be elaborated and enforced. Then the Earth will be able to sustain our increasing numbers and our most basic needs. We in turn will be able to preserve and restore what is left of the life and beauty of the natural world. And Nature's nemesis will become our apotheosis. The prevailing dominionistic and adversarial attitude toward Nature will be transformed into a more creative, mutually enhancing relationship. Progress for our species will be measured by a qualitatively different set of criteria that reflect a more mature, reverential respect for all life. This is the bioethical basis for a humane and sustainable world economy, as well as for a community of hope and fulfillment.

Toward a Sympatric Reverence for All Life

The source of human morality and ethical sensibility is surely in our very nature, in our genetic makeup. We have been honed by millions of years of evolution into a cooperative, communicative and increasingly empathetic primate. Natural selection processes favoring group survival reinforced these human attributes and helped lay the foundation for the establishment of more complex social groups and subsequent regional principalities and nation-states. Many of the teachings, parables, and commandments of the world's major religions are concerned with harmonizing human relationships and communities, and in encouraging individuals and communities to exploit animals and natural resources with respect and care. Animal and environmental protection are evident concerns that are deeply embedded, along with social justice, in the religious history and civilizing process of our species. It is evident that such morality and ethical awareness evolved along complementary lines in various civilizations and cultures separated from each other in both time and space.

But today, as some social critics speak of the end of history, the death of God and of Nature, we find that the moral fabric that once held communities together is falling apart. Violent confrontations between different socioeconomic, religious, cultural and racial groups and classes reveal a side of human nature that emerges when moral constraint and empathic and ethical sensibility are lacking. This lack of restraint and sensibility is reaching epidemic proportions and is symptomatic of a pathological state of mind. What is especially disturbing is that those positive attributes of human nature that we acquired over the millennia to enable us to survive, prosper, and resolve conflicts cooperatively can be so easily supplanted—supplanted by attitudes and values that arise from a pathological state of mind that is primitively programmed to ensure the survival of the individual regardless of the survival of others, even close kin.

This unbridled egotism has its neurochemical roots in the pre-human subconscious which comes to the surface and affects how people think, feel, and act, when the moral constraints, empathic, and ethical components of rational human consciousness and compassionate conduct are torn away by circumstances such as poverty and famine, if not by choice, then chance. For example, the tragic consequences of disenfranchised indigenous peoples so well documented in Africa by Colin Turnbull were repeated in the genocidal conflagration in Rwanda in 1994.[3] What some refer to as the beastly or demonic side of human nature should be seen as a primitive, indeed archaic, survival mechanism that is activated under certain circumstances. These circumstances are precisely those that are created by economic and environmental crises that arise from the unfettering of those moral constraints and empathic and ethical sensibilities that normally keep the powerful will-to-be of the human spirit or ethos contained.

When unfettered, this will-to-be—what neo-Freudians and others often demean as the Id—is an extremely potent force, especially in the human species. In other animals, the Id is more fettered by instinctual processes and by psychophysical niche adaptations and limitations. Co-adapted and co-evolved species are thus self-restrictive or are biologically constrained, and mutually enhancing. In humans, this will-to-be must be consciously fettered by the guiding and tempering influences of bioethical rationalism and a heart-filled sympatric reverence for life.

When the human Id or will-to-be is directed only to serve the self or ego, or some deified ideology like fascism, tribalism, racism or industrialism, this pathology becomes epidemic and great suffering and desecration ensue. But when this will is consciously directed in reverence and resonance with our ultimate point of reference and ground of being—the Universal Self—the good of all life is enhanced. As Spinoza observed, "We are all modes in the Body of Being." Fettered by reason and compassion, the Id enables the human ethos or spirit to find its rightful planetary niche, its cosmic place as a co-creative, co-evolving being, along with myriad other life forms. We must not forget our sacred connections, origin, and purpose, and we must not be distracted by our cleverly fabricated delusions of illimitable power and instrumental knowledge.

How we come to live in harmony with this state of mind is our present evolutionary and survival challenge. The state of the world, of the places we inhabit, exploit, glorify, or desecrate, mirrors our collective state of mind. We will not, therefore, make this world a better place until our state of mind is focused on consciously directing our will-to-be toward a more sympatric relationship with each other and the entire life community of planet Earth. Until we realize the wisdom of the Sabbath, of taking time out to reaffirm and repair our connections and responsibilities toward all that live in the places we inhabit and exploit, the suffering and desecration that we cause will consume all that was the glory of Creation—the life and beauty of this planet.

Evil Means—Vivisection: Contested Terrain, Beastly Questions[4]

As a former animal researcher myself, I am no stranger to altruistic rationalizations of vivisection (experimentation on live animals). One rationalization is that "the suffering of the few for the benefit of the many is justifiable." Yet the only true animal model for human disease is man himself. Vivisection is in itself symptomatic of a diseased attitude that no amount of animal research and suffering and killing will ever cure. Metaphysically, our "dis-eased" condition in part reflects the state of mind and the social consensus that accepts the killing of human beings in terrorism and war, on the one hand and vivisection, the infliction of suffering and killing of animals, on the other.

The similarity between terrorism and vivisection warrants some reflection. Both entail a deliberate, indeed calculated transgression of the doctrine of ahimsa and of the Golden Rule, inflicting suffering and death to achieve some purportedly greater good. Just as the antiterrorist does violence in the name of peace, so too the vivisector does violence in the name of medical progress. Both are accepted by one segment of society or another, if not as altruistic acts to further some greater good, then as necessary evils. Even though the deliberate infliction of harm is an evil in itself, in both instances the absolute ethic of ahimsa, avoiding harm to all living entities and to the natural environment, is overridden by situational ethics that condone evil means for purportedly just ends. The circle of violence is completed when, like the antiterrorist, the antivivisectionist engages in threats and acts of terrorism (as distinct from destroying property and assuming protective custody of animals) against biomedical researchers in the name of animal liberation.

There are similarities between terrorism and vivisection. In terrorism, as in vivisection, there are innocent victims—the animals that have been variously bred and captured and who are regarded as inferior. In the secular, if not profane, world of medical science, these animals are not generally seen as manifestations of divine Creation, as belonging to God and therefore "ours" only in sacred trust, to be treated with respect and love. Under the perceived threat of disease, suffering, death, and loss of our loved ones, we devalue the lives of animals, violating the sanctity of life by valuing our own over our fellow creatures. It is this attitude of mind which fears death and suffering and which, by objectifying certain other sentient fellow beings, human or nonhuman, we become empathetically disconnected from them and see them as inferior, less important than we are.

This disconnectedness can fulminate into sociopathic, biopathic, and zoopathic behaviors that collectively lead to the destruction of Earth. Collective biopathic behavior is exemplified by industrialism's destruction of the natural environment, and by those values that place material gain over ecological sensibility—and which inevitably lead to economic instability and environmental disease. Institutionalized zoopathic behavior, as in vivisection, factory farming, and the wholesale harvesting of wildlife species, underlies the cultural and ethical disintegration of industrial society and its ultimate nemesis by way of materialism and secular humanism.

Recent developments in genetic engineering biotechnology, which entails considerable animal experimentation and suffering, will only serve to hasten the demise of industrial society if this biotechnology is not integrated with a more healthful, ecologically sound, and humanely sustainable agriculture. It must be integrated, too, with a more holistic, environmentally, and ecologically oriented medical paradigm. The modern medical paradigm is inadequate insofar as it barely addresses the environmental health problems associated with industrial pollution and agrichemical poisons. It will remain inadequate in the absence of appropriate political involvement. Applying new developments in medical science and genetic engineering biotechnology to help us adapt to an increasingly pathogenic, disease-enhancing environment is a massive denial and a waste of public resources and good minds.

It is from this perspective that vivisection should be abolished. Although it has contributed to the advancement of medical knowledge, it now stands in the way of any further significant medical progress that will be of benefit to humanity in these times of environmental degradation and cultural crisis. Indeed, vivisection contributes to the failure of medical science to prevent human disease and alleviate human sickness and suffering. So often this is a consequence of harmful treatment side-effects—of a false reductionistic and mechanistic orientation, which arises out of false hopes and promises of recovery based on animal models that for many human diseases are of poor fidelity and are scientifically invalid. The aphorism "physician do no harm" has been preempted by the reductionistic and mechanistic approach of drug-dependent allopathic medicine, and with it, dependence upon vivisection and the justification of laboratory animal experimentation for the good of society.

A reverential attitude of heart and mind automatically precludes vivisection because one's own self is felt to be of equal significance and sanctity as the life of any other living being. Animal research is then restricted to the care and study of animals that are already sick and injured, for their benefit and the possible future benefit of their kind. Coincidental but nonetheless significant benefits in the advancement of human health would arise naturally from the knowledge and skill gained in treating such animals.

A reverential attitude toward all life would preclude the use of normal, healthy animals in psychological research, in stress, pain, and trauma studies, in toxicity and cosmetics testing, and in military weapon and biological warfare testing. Genetic engineering and selective breeding of sick and mutant creatures would also be unthinkable. But reproductive, nutritional, genetic, and other studies of endangered species to facilitate their viability and reintroduction into restored and protected natural habits would be accepted, since the animals themselves would be the primary beneficiaries of such research.

Research on farm animals to enhance their productivity, efficiency, and adaptability to factory farming conditions would be something of the past, since the ethic of ahimsa mandates vegetarianism, or at least a dramatic decrease of consumption of animal products by all industrialized nations. Concern for wildlife, both terrestrial and aquatic, necessitates such a shift toward vegetarianism and a sustainable agriculture to protect natural habitats and wild species displaced and exterminated by overfishing and by the conversion of natural habitats to provide feed for farm animals. Veterinarians need to terminate research on the diseases of other commercially exploited animal species, from "exotic" pets to racehorses and ranch-raised wildlife, including fish, alligators, turtles, deer, and fur-bearing mammals. An enlightened, compassionate society would not be involved in such forms of ethically questionable animal propagation and exploitation in the first place.

We cannot expect the terrorist to kill the "terrorist within" or lay down the gun without pressure from the public. Nor can we expect the dedicated vivisector to lay down the needle, scalpel, electrode, or laser on his own initiative. Without increasing public awareness, legislative pressure, and consumer boycotts of all companies and industries whose products and services cause animal suffering, coupled with educational reforms that foster an empathetic and compassionate respect and reverence for all life, the unjust and unnecessary suffering of animals in biomedical laboratories around the world will continue.

The abolition of vivisection should be seen as part of the healing process of humanity. We have become so disconnected from reality that we are still at war with ourselves and condone the killing of human beings in the name of justice, security, and peace, while the planet we inhabit is dying. This is not an overstatement. To contend otherwise, or to believe that more animal research and new technological and legislative correctives will suffice, is a denial of reality. And to claim, simplistically, that the antivivisectionist is sentimentally misguided and cares more for animals than for people is a gross injustice to those whose ethical sensibilities are clearly beyond their critics' comprehension.

Denial and fear, along with ignorance, arrogance, and greed are great obstacles to human progress. The end of vivisection could herald a new beginning where all policies and decision making at the personal, corporate, and political levels are based upon the three principles of a humane, planetary society, namely: obedience to the Golden Rule, ahimsa, and reverence for all life. With this, the integrity of Creation and the future of humanity, itself, may be better assured. In the final analysis, animal liberation and human liberation are one and the same. They are consonant with and complement the ethics and morality of a truly civilized society.

The doctrine of ahimsa, of avoiding harm to living beings, has an ancient history. The next chapter revisits this doctrine. It is one of the most important and fundamental bioethical principles and needs to be embraced by all today.

CHAPTER 12

Ahimsa (Noninjury) Revisited

The ancient Sanskrit word *ahimsa*, meaning noninjury, is used for the doctrine of refraining from harming others. It is the central teaching of Jainism, Hinduism, and Buddhism. As an ethical principle, we find it in the Judeo-Christian concept of the Golden Rule that holds that we should not do to others what we would not have them do to us. And it is implicit in the medical maxim "physicians should do no harm."

The principle of ahimsa has a long history. It originated with the Jain religion of India several centuries before the common era. The active principle of ahimsa is termed *daya*, meaning compassion, empathy, and charity. Ahimsa implies *jiva-daya*, actively caring for and sharing with all living beings, tending, protecting and serving them. It entails universal forgiveness (*Kshama*), freedom from fear (*abhaya*), and universal friendliness (*maitri*).[1]

Another ancient Jain scriptural aphorism, *parasparapagraho jivanam*, meaning that all life is bound together by mutual support and interdependence (which the science of ecology has subsequently confirmed), gives an added dimension to ahimsa—namely that one avoids harm to the natural order and balance of Nature by respecting the universal interdependence of all life. In order to live this way, people and the populace alike must strive for equanimity (*samyaktva*) toward both animate (*jiva*) and inanimate (*ajiva*) objects and substances. An attitude of give and take, and of live and let live is thus encouraged.

The Jain theory of knowledge, *anekantavada*, holds that reality has manifold aspects with an infinity of viewpoints. This is the basis for the doctrine of *syadvada*, or relativity, which states that truth is relative to different viewpoints. This means that no one has a monopoly on truth, and that we do violence to others when we do not

175

listen to their views and discount their concerns. Absolute truth cannot be gained from any particular viewpoint, alone, because absolute truth is the sum total of all the different viewpoints that make up the universe. Anekantavada is, therefore, the essence of planetary democracy, since it does not look upon the world from an egocentric, ethnocentric, or anthropocentric perspective.

The doctrine of ahimsa is a call to ethical action. This active principle was termed *satyagraha* by Mohandas Gandhi—the power of compassionate action. Gandhi, one of the great proponents of the doctrine of ahimsa, clarified this ideal as follows:

> Strictly speaking, no activity and no industry is possible without a certain amount of violence, no matter how little. Even the very process of living is impossible without a certain amount of violence. What we have to do is to minimize it to the greatest extent possible. Indeed the very word non-violence, a negative word, means that it is an effort to abandon the violence that is inevitable in life. Therefore, whoever believes in *ahimsa* will engage himself in occupations that involve the least possible violence.[2]

Actions that entail encouragement of life need to be as carefully considered as those that entail the deliberate, unavoidable taking or harming of life. This is because our most altruistic actions can have harmful consequences to others if we do not follow the doctrine of ahimsa or the Golden Rule. While we cannot live by the Golden Rule as an absolute, we absolutely must consider the Golden Rule prior to deciding upon any action. We should be mindful of the differences between the unavoidable, natural, and pervasive suffering we see in Nature and the often avoidable human-caused suffering, over which we do have considerable control.

The only absolute principle is to have reverence and respect for all life. This does not preclude the unavoidable harming or taking of life, since we cannot, in these times, live by the absolute ethic of ahimsa or the Golden Rule. We do, however, have the absolute responsibility of governing ourselves according to these ethical principles for the good of all. For example, we regrettably must accept the humane destruction of "surplus" elephants to help preserve herd and habitat when there are no alternative solutions available, such as a method of birth control or more elephant habitat. Likewise, animal shelters around the world engage in euthanizing millions of homeless cats and dogs. But in all such instances, humane alternatives must be sought for future application, so as to avert the continuation of situations and circumstances incompatible with the doctrine of ahimsa.

The doctrine of ahimsa encompasses both human and nonhuman life. It also embraces non-living entities such as lakes, swamps, and all natural ecosystems that can be harmed by various human activities, that in turn may harm the animal and plant communities therein. Philosopher Knut A. Jacobsen has written a very relevant article on this topic, which he summarizes as follows:

> The principle of non-injury toward all living beings (*ahimsa*) in India was originally a rule restraining human interaction with the natural environment. I compare two discourses on the relationship between humans and the natural environment in ancient India: The discourse of the priestly sacrificial cult and the discourse of the renunciants. In the sacrificial cult, all living beings were conceptualized as food. The renunciants opposed this conception and favored the ethics of non-injury toward all beings (plants, animals, etc.), which meant that no living being should be food for another. The first represented an ethics modeled on the power that the eater has over the eaten while the second attempted to overturn this food chain ethics. The ethics of non-injury ascribed ultimate value to every individual living being. As a critique of the individualistic ethics of non-injury, a holistic ethics was developed that prescribed the unselfish performance of one's duties for the sake of the functioning of the natural system. Vegetarianism became a popular adaptation of the ethics of non-injury. These dramatic changes in ethics in ancient India are suggestive for the possibility of dramatic changes in environmental ethics today.[3]

Some philosophers reason that since some animal species are more sensitive and intelligent than "lower" life forms, they have more "intrinsic" value. Hence they believe these animals (like elephants) should receive more respect and protection than "lower" life forms, like worms and insects. I believe this line of thinking is anthropocentric and speciesist. So-called "lower" life forms in healthy, natural ecosystems have great extrinsic value—they contribute to helping maintain the functional integrity of ecosystems, the balance of Nature. For example, earthworms are soil makers, and various insects pollinate plants. Despite their relatively low degree of

sentience, these and other "lowly" creatures play a far more significant role than most humans in their contribution to the well-being of
the natural world.

The doctrine of non-injury does not limit respect and compassion
to living entities based upon their degree of sentience, but includes
non-sentient living ecosystems within the scope of moral consideration and empathic concern. Critics might argue that because it is so
all-embracing, the doctrine of ahimsa is an impractical and unrealistic ideal. Yet by virtue of its illimitable scope, it takes us beyond the
polemicizing dualities of animal versus human rights and human interests versus environmental protection and nature conservation. It
is surely from such an all-encompassing ethical sensibility that we
can best consider, rationally and sensitively, the rights and interests
of the entire life community of the planet. The doctrine of ahimsa is
the cornerstone of a just, humane, and sustainable society. It is also
enlightened self-interest because when we harm others, including the
environment, we inevitably harm ourselves.

This latter point leads us to a related principle of Eastern religious teachings, namely the law of *karma*. One's destiny is influenced by one's thoughts, words, and actions. (What goes around,
comes around.) The law of karma recognizes that good will ultimately comes to those who endeavor as best they can to live according to the doctrine of ahimsa. But this is no easy task when we are
born into a culture where social discord and violence are endemic
and contagious, where cruelty toward animals is condoned and institutionalized, and where the destruction of the natural world is
economically rationalized and industrially sanctioned. It takes great
courage, commitment, and vigilance to live in accord with the doctrine of ahimsa in a culture whose values are antithetical to this
compassionate ethic of non-injury. Yet the more we can disengage
our lives from those forces that are responsible for so much suffering
and destruction in the world today, and still enjoy productive and
meaningful lives, the more society will change and become more humane, socially just, and environmentally sustainable. For example,
we can disengage as consumers from supporting cruel factory farming systems, by not purchasing various animal products from such
farms. We can also support organic farmers by selectively purchasing their produce and buying cosmetics and other consumables that
have not been tested for consumer safety on animals, and which contain no ingredients of animal origin.[4]

While there seems to be no alternative but to be part of a culture
that is the antithesis of ahimsa, as predominantly urban-dwelling

consumers, often employed in industry-related businesses that value economic growth and material profit over environmental and animal protection (and even over consumer protection and worker safety), there are some choices that we are still free to make in accordance with the doctrine of ahimsa. We can choose to eat less meat or to become a vegetarian. We can choose to have a small, fuel-efficient car, to recycle household and office trash, to buy "cruelty-free" toiletries that have not been tested on animals. The choices are many, once we become more vigilant, informed, and dedicated to live as best we can in gentler ways that cause less harm to others and the natural world. And every choice that we make is a vote that will make a difference.

Certainly we must exploit life in order to sustain our own. In natural ecosystems, one life supports another. The food chain reveals how interdependent each life form is, and how each life gives as much as it takes so that ecological balance is preserved and the system remains sustainable and self-renewing. We have been slow, though, to apply these scientific findings and natural laws to modern agriculture and other industries. It is noteworthy that in every healthy, balanced ecosystem, every life-form plays an integral role. Even if it takes another's life, it still causes more good than harm to the life community within that system. Since the human species is less constrained than other creatures and has the powers of free will and dominion to act outside of natural law, we must, for the good of the whole and for our own good, apply the guiding principle of ahimsa to help ensure that when we exercise these powers, we cause more good than harm to the life community.

The cultural assimilation of the doctrine of ahimsa is the hallmark of a truly humane society. We have much work to do to help lay this foundation for such a society. That we will never enjoy it in this lifetime is no reason for us not to begin to build our own lives around this doctrine, for the good of all and for generations to come.

Animals have served many human needs over hundreds of thousands of years. They have variously provided us with food, shelter, social status, clothing, labor (as for draft work, pulling ploughs, carts and sleds), and have served loyally as companions and guards of home and livestock. Many animals, especially dogs, have heroically saved their human companions from accidental drowning and fire, live burial beneath avalanches, and suffering under the burden of loneliness and depression.

Our demands upon animals have increased rather than decreased over the millennia as human society has become more industrialized, if not actually more civilized. The following examples of widespread animal cruelty and suffering clearly illustrate that our power of dominion over animals is abusive, and that without concerted effort, contemporary society will continue its ethical and spiritual decline and suffer the consequences:

- The meat, eggs, and dairy products we consume come mainly from animals raised in cruel factory farms where they are stressed and made susceptible to disease by extreme overcrowding in cages or pens that also leave them unable to walk or turn around. Until these systems are changed to provide animals with environments that better meet their physical and psychological needs, we owe it to them to fight such inhumane production methods. We can do this by eating less or no animal products from factory farms, by selectively purchasing produce from farmers and ranchers who have adopted less intensive and more humane methods of livestock and poultry production, or by becoming vegetarians.

- Many consumables, from household cleaners to cosmetics and other toiletries, have been safety-tested on animals. These laboratory tests often result in great suffering. Concerned consumers purchase products that are either clearly marked as not having been tested on animals, or buy "tried and true" brands rather than "new and improved" products that have most likely been tested on animals. The suffering of animals for such trivial ends cannot be justified. Many compassionate consumers go one step further and boycott any products that contain animal ingredients, like perfumes that contain animal musk oil and soaps and cosmetics that contain animal fat (tallow) and oils.

- Items of adornment, from fur coats to many leather goods and jewelry made from various animal products are avoided by those who care for animals. Furs come from wild animals that are caught and suffer great anguish in steel jaw traps and snares, or from wild animals raised in small cages on fur farms where conditions are no better than on cruel factory farms. Other animal products come from rare and endangered wild animals that are killed merely for their ivory, skins, or other body parts, to be used to make jewelry and other accessories—and even folk medicine in the Orient.

- Various animals kept as companions or pets come from a commercial pet trade that all caring people boycott by adopting animals instead from the local animal shelter. Many purebred puppies for sale in pet stores come from "puppy mill" factory farms that are often as cruelly deplorable as livestock and poultry farm factories. Other more "exotic" pets, like parrots and other creatures caught in the wild, suffer high mortalities before they ever reach the pet shop. And they don't make good pets, since they have not been bred to adapt to captivity and to domestication.

- Animals also suffer in the name of sport and entertainment. Such activities and events as trophy hunting and sport fishing, rodeos, horse races, animal circuses, and roadside zoos, do not enjoy the support of those who have a vestige of empathy for wild or tame animals.

These examples affirm my contention that until all such abuses cease, the law of karma will ensure that society will continue to be dysfunctional and violent. Also, as society continues to treat animals and the rest of Creation with cruel indifference, we will continue to bring ecological and socio-economic catastrophes upon ourselves and upon generations to come. As consumers we can be empowered by the doctrine of ahimsa to choose wisely and with compassion. By so doing, we help undermine the economic incentives that are the primary reasons for the continuation of so much animal cruelty and suffering. As voting citizens, we can support local, state, and federal environmental and animal protection legislation, and push for better enforcement of such laws.

Developmental and Educational Considerations

Our self-interest can be so self-centered that it leads us to have no regard for Nature except as a resource. Our self-interest can also lead us to have little or no compassion for animals and respect for the inherent value of all of Earth's Creation. The evolution of species is deeply rooted in self-interest, from self-preservation to self-perpetuation. The evolutionary success of the human species, however, is turning into a scenario of tragic failure. This is in part because our impact upon the planet has expanded globally, but this expansion has not been accompanied by a comparable expansion of our sense of self and responsibility as a planetary species. We are the earth, insofar as our selfhood or being is connected historically, ecologically, and spiritually with the entire life community of the Earth.

This symbiotic life community, as Thomas Berry has proposed, is built upon a communion of subjects rather than upon a collection of objects. Within this community we find a sacrificial dimension where life gives to life in order to sustain the entire community. While we are physically, and to a degree unconsciously, connected with this community—with the rocks, trees, waters, air, the food we eat, and so forth—we can become consciously connected with the life community through our ability to empathize, to put ourselves in another's place. Empathy connects and universalizes the self with the suffering, joy, wonder, and mystery of all life. Without empathy, we become disconnected. The result is that subjects that deserve empathic communion and celebration are turned into objects that we variously demean and exploit. In the process, we do no less to ourselves: and as we empty the cosmos of *interiority*, of subjectivity, intrinsic value, and significance, we do the same to ourselves and to each other.

Healthy children have a natural capacity to empathize, a capacity that parents and others must nurture. But too often this essential attribute of our humanity is crushed, if not by parents, then by the values and attitudes children acquire through education and even religious instruction. Ethical sensibility arises naturally from empathic sensitivity. The absence of empathy means the absence of ethical sensibility, which in turn necessitates the imposition of law and order and often blind (unfeeling) obedience to moral codes.

In order to help ensure that the ability to empathize would become integrated with the ethical and spiritual precepts of the community, many preindustrial civilizations carefully nurtured and educated their children, especially through example and initiation rituals. Initiation rituals were designed to reintegrate the developing sense of self (our adolescent egos) with both the unconscious side of our natures and with the ecos or natural world and life community around us. Unintegrated, the adolescent human ego is a terribly selfish and potentially destructive force.

When the Earth is poisoned and its ecology dysfunctional, human health and a functional society are unattainable ideals. To heal ourselves we must heal the planet, and to heal the planet we must heal ourselves. But nothing will be well until we show respect and compassion toward our fellow creatures. Without this, injustice and inhumanity will continue to ravage every human community around the world that sees itself as superior to and separate from the rest of Creation. The ancient doctrine of ahimsa should be the unifying principle that links animal and environmental rights and protection with human rights and interests. No philosophy of bioethics, no constitution, religion, industrial economy, or technology is acceptable without this unifying humane principle.

Ahimsa, Reincarnation, and Transgenerational Equity

The Jain principle of ahimsa is the rational basis for all altruistic actions, institutions, and communities. It is the key to human health and fulfillment across generations. It establishes an objective and empathic bridge of responsibility for all life. This enables us to extend active compassion toward all life. We become especially mindful of the invisible creature-souls in the air we breathe, the water we drink, and in the soil that feeds us all.

For Jains, Hindus, and Buddhists, belief in reincarnation strengthens the sense of transgenerational continuity and responsibility. The Iroquois and other native Americans simply called this *The Law of Seven Generations*. This concept is difficult to find in the scriptures of Judaism and Christianity.[5] And it is not evident as an active, organizing, and guiding principle in contemporary society. If everyone believed in reincarnation and had reverence for all our generations and relations, past, present, and future, this world would be a different place.[6] Contemporary social observers call this process of establishing a temporal dimension to our ethical sensibility, transgenerational equity.

All sustainable societies in the past have lived in this reverential dimension, perhaps because they knew that a lack of transgenerational respect leads to an increasingly selfish, competitive, and destructive way of life, as well as to a dysfunctional society and economy. Even if the notion of reincarnation is unacceptable, we should reflect upon the sage advice of the Prophet Muhammad that one should "live in this world as if you were going to live forever; prepare for the next world as if you were going to die tomorrow." As the reincarnate Dalai Lama once proclaimed to me, "If you must be selfish, at least be altruistic."

Clearly, any definition of altruism must include the principle of transgenerational equity as well as the ethic of respect and reverence for the living community of this planet, past, present, and future. Such a temporal view of equity is surely the cornerstone for a just and humane society, as it is the rational basis for a sustainable agriculture and for all human activities that affect the life community of the Earth. When was the first breath of life and to what final end or greater purpose was that life created? Some people doubt there is any divine purpose or plan. They believe that there is no sacred dimension to the spotted owl, or monarch butterfly. Everything becomes extinct—species, individuals, even ecosystems—due to natural disaster and what is erroneously called evolution. Only the human soul, many contend, is immortal and thus has some superior purpose over the rest of Earth's Creation.

But was it only for the sake of Homo sapiens that the first breath of life was ever made? What creative intelligence, or random process that became self-organizing, created us? Was everything else simply a means to the creative end of man, perfected in the image of some God-Creator? Or is everything that was, is, and shall be, the manifestation of ineffable divine mystery? Was that first breath mere chance, that great outpouring and inflowing *prana* of energies coinhering, multiplying, and diversifying into myriad patterns and intelligent embodiments of life—was that without purpose? If there is sacred conception and purpose, then there is a sacred end. The means for the realization of the sacred are no less sacred. Therefore, to those who reason that man is the realized image of God, then the means whereby that image was made possible are no less sacred. And those means are the oceans, the rivers, the forests, and the myriad creatures and plants before us, which now sustain us in body and spirit.

For the nihilist, atheist, agnostic, and secular humanist, the concept of self-realization or God-realization is anathema. Even so, for them acceptance of bioethics is enlightened self-interest. To regard all nonhuman life as having been created primarily for human ends, or to contend that life is ultimately meaningless because it ends in death or extinction, is to demean the very significance of human existence. As a species that knows that it knows—that is self-reflexive—we are conscious participants in a cosmic process, the totality of which is beyond objective comprehension. How can we not then acknowledge the phenomenal mystery and wonder of life? How can we not then respect and celebrate its sanctity? But without justice, as the next chapter considers, respect, ahimsa, and compassion are simply words without action.

CHAPTER 13

Justice and the Communal Good

The Golden Rule is the keystone of common law and community relations. Justice is a basic bioethical principle that has been short-changed through a singular focus on human rights and interests. Extending justice to include regard for animals and Nature effectively accomplishes the long overdue paradigm shift from anthropocentrism to biocentrism.

The environmental and animal rights movements have been major catalysts in bringing about this societal change in perception, arguing that it is unjust to treat animals inhumanely and to degrade the environment. The eco-justice concept has been adopted by the U.S. Presbyterian Church and the World Council of Churches, among others. Just treatment of animals (zoö-justice) is also gaining acceptance as a result of efforts to document and acknowledge connections between animal cruelty and crimes of violence, and indifference toward institutionalized animal cruelty and exploitation and a violent, uncaring society. A sense of justice arises spontaneously when the basic bioethical principles of compassion, reverential respect, and ahimsa become central to our personal and professional lives and to community, government, and the corporate world.

In the absence of compassion, respect, and ahimsa, justice must be imposed. This is done preferably from within the community. But as communities disintegrate, and as injustice becomes the commonplace currency of human relationships with their own species and with other animals and Nature, justice must be enforced from without. A judiciary system thus evolves out of necessity as something external to the ethos of a community that is incapable of self-governance. The more dysfunctional the community, the more litigious it becomes. And the more authoritarian and dysfunctional the judicial

185

system becomes, the more the community suffers the violent consequences of what has become its own nemesis.

When responsible authority is passed from communities and corporations to centralized state and federal government, we have a disempowered populace and irresponsible corporations. These corporations fight government control that may limit their profits and monopolies, and have undue influence over legislation and the judicial system. Lacking such influence, the disempowered populace trusts neither politicians nor corporations. Yet, ironically, the public response to corporate activities and products that harm the environment and consumers—as exemplified by the frequent epidemics in the U.S. of E. coli food poisoning from contaminated hamburgers and other foods—is to demand more government oversight, inspectors, new regulations, etc. There is no call for the system and values that created these problems in the first place to be changed. Essentially it's business as usual while the public pays for more government lip service instead of achieving any significant increase in responsible corporate ethics and accountability.

The strong alliance between corporations and government is exemplified by the governments of so-called "industrial democracies" touting the Gross National Product (GNP) as the cardinal index of progress and the social good. Yet most current economic development programs and increased industrial productivity efforts invariably have adverse social and environmental consequences. We clearly need better indices of progress that incorporate bioethical principles and address such interconnected concerns as environmental quality, biodiversity, social well-being, and the effectiveness of environmental, animal protection and human rights legislation.

Yet these very concerns are seen as obstacles to corporate interests and potential market barriers to international trade. Corporations even have their employees, like the loggers in the Northwest and the factory-farm contractors and managers in the Midwest, believing such fiction. Yet in the final analysis, neither justice nor the social good can be served or assured by such materialistic criteria as the GNP, or by the limited goals of economic development and industrial growth. The strength of social economy can be measured by the strength of mutually cooperative and enhancing community relationships and by its members having secure, satisfying and sustainable livelihoods. Today's global market economy is undermining the very basis of its own sustainability by destroying the social economies of every nation-state, and squandering nonrenewable natural resources (see Fig. 13.1).

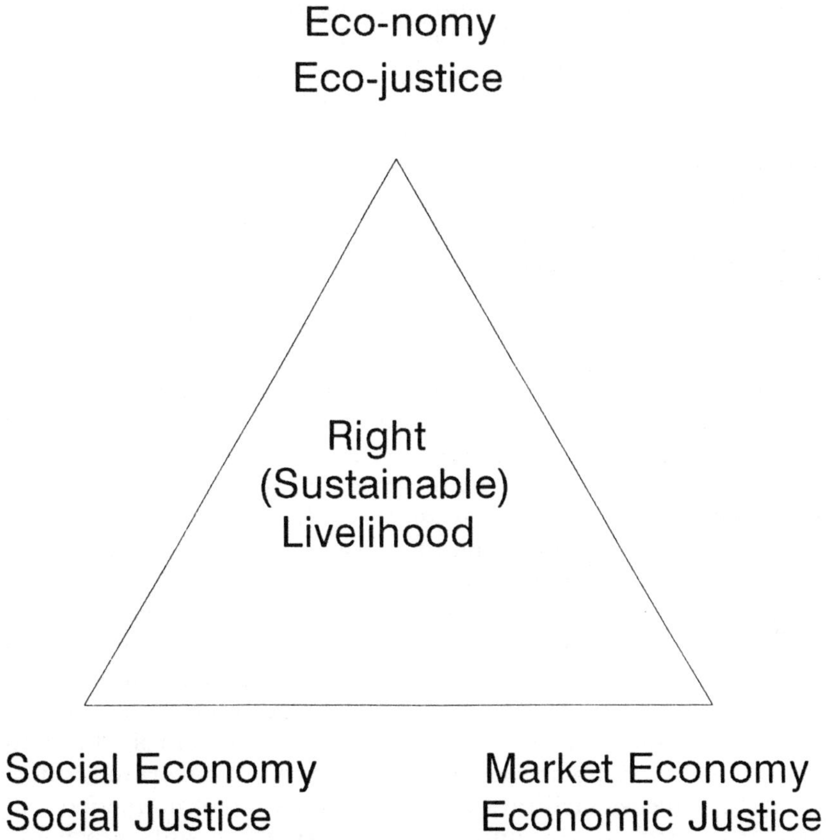

Eco-nomy
Eco-justice

Right
(Sustainable)
Livelihood

Social Economy
Social Justice

Market Economy
Economic Justice

Fig. 13.1. The three interdependent components of justice, when integrated, lead naturally to a sustainable community.

Implementing Sustainability: Values, Principles and Justice

In his provocative book *Goatwalking: A Guide to Wildland Living: A Quest for the Peaceable Kingdom,* Jim Corbett establishes himself as a contemporary Thoreau, observing that, "The problem is not that modern man wants so much but that he aspires to so little. . . . Right livelihood *is* practice that reflection knows as communion. . . . Only through right livelihood can reflection see the virtue—the life-enhancing power—of the cocreative community's masterless morality."[1] Corbett's central concern is to develop a biocentric ethic to counter the ecologically and socially destructive consequences of industrialism, capitalism, and atomized anthropocentrism. In this spiritually and politically challenging book, Corbett—philosopher-goatwalker and co-founder of the Sanctuary Movement (which helped "illegal" Salvadoran and Guatemalan refugees immigrate to the U.S. across its southwestern borders)—contends that, "Whatever the relative merits of trying to engineer human harmony with the rest of life on earth by means of government management (instead of trying to cultivate it by community covenant), a land ethic that seeks to extend basic rights to the land and its life must take the way of covenanting. If the land ethic visualized by Aldo Leopold is to emerge, the private ownership of land must be hallowed by community covenants, not absorbed into centralized state management."[2]

With ever increasing numbers of people living in urban centers and dormitory suburbs from coast to coast and nation to nation, the very notion of any kind of integrated working community sufficiently organized to establish any hallowing community covenants seems ever more remote. And yet this ideal is conceivable and achievable as neighborhoods form their own self-governing bodies and in the process facilitate the decentralization of municipal and state authority, control and responsibility. Covenants are being made by neighborhood, church, and school communities to protect and restore parks and other urban wildlife habitat, to recycle household wastes, to provide for the homeless via donations to food and clothing banks, and to support urban market gardens and local farmers and horticulturalists. Indeed, community supported agriculture (CSA), especially of organic and humane sustainable farming systems, is one of the most important covenants linking urbanites with rural communities that want to hallow the land and put the land ethic into practice. With the support of urban consumers, CSA demonstrates that hallowing community covenants are achievable, and politically empowering.

The restoration of the social economy and ecosystems is the basis for sustainable economic development (see Fig. 13.1). This is the praxis of economic, social, and eco-justice, which is enlightened self-interest for the rich and for multinational corporations. David Korten, economist from the People-Centered Development Forum, provides an incisive critique of nonsustainable development:

> Nearly fifty years of international development effort have focused public policy and resources on efforts to accelerate the growth of monetized economies. These efforts have achieved a five-fold increase in global GNP since 1950. Yet unemployment, poverty, and inequality continue to increase, the social fabric of family and community is disintegrating, and the ability of the ecosystem to support human life is being destroyed—all at accelerating rates. Left without adequate opportunities for productive employment, a major portion of humanity is marginalized from the mainstream social, political and economic processes of the societies in which they live, and more than a billion people are consigned to lives of abject poverty. The tragic irony is that while a wide range of essential needs go unmet, hundreds of millions of people have been forced into unproductive idleness or meaningless work.[3]

Financial Aid Charade

The old adage "neither a borrower nor a lender be" should be applied to GATT as "neither an importer nor an exporter be." Unless the financial security of a country and every community therein can be guaranteed, the risks and costs of pinning economic growth and quality of life on exporting various agricultural and other commodities will far outweigh any benefits. The frantic U.S. government's bail out of Mexico's plummeting peso in February 1995 was done in part to protect U.S. businesses by helping Mexico pay for imported produce from the U.S.

It would seem that the financial aid given to Mexico is in part an indirect subsidy or price support paid for by the U.S. taxpayer, that we are told benefits U.S. companies based in Mexico. But in actuality the benefit goes to a few multinational corporations like ConAgra and Cargill that enjoy a market monopoly. These companies have the power to make one country or block of countries give loans to developing countries and to those experiencing a fiduciary crisis, in order to maintain their hegemony. Their global influence is as pervasive as it is pernicious. Rather than relying on the military to expand their empires, they use money as a weapon. Loans from agencies like the World Bank and International Monetary Fund are used to further the multilateral interests of these corporations, who use the loan moneys to buy machinery, seeds, pesticides, etc. Through GATT and WTO, this unethical business will enjoy legal protection.

But all to what end? Disappearing forests, eroding soils, falling water tables and collapsing fisheries are now combining to threatened global climatic and socio-economic chaos. This is the legacy of the commoditization of life and Nature's resources, of an industrialism that has made the natural world poorer by making our wants many. And it is the reason why the Green Revolution failed from the perspective of third world farmers and communities (see Addendum Ch. 7), although the World Bank and Food and Agriculture Organization would have us believe otherwise. We have certainly changed the economic structure of the world and become entangled therein to an extensive degree. Henry David Thoreau probably envisioned this over a century ago when he wrote, "I make myself rich by making my wants few."

Corbett concludes, "As life becomes reflective, we must choose either to live demonically by trying to possess the world, or prophetically by actively participating in creation."[4] Through obedience to the Golden Rule, and by extending the principle of justice to embrace all sentient life, we may indeed discover that living prophetically and establishing a hallowing covenant to give just consideration to all creatures and Creation is the highest form of enlightened self-interest.

Corporate Justice and Responsibility

The basic values of a consumer society addicted to materialism center around self-gratification and the means, often selfishly competitive, whereby the pleasure principle can be satisfied with speed and certainty. There are few, if any, viable values related to strengthening the spirit of community and cooperatively enhancing the social economy, because both have been virtually eliminated in the overdeveloped industrial world by the influence and growth of corporations. Corporate activities that dismember communities and obliterate family farming in rural America have been aided by local and federal governments, and by land grant colleges under the State

University system, all in the name of progress and efficiency. These activities, in actuality, have meant economic growth and industrial expansion for monopolistic corporations that have grown to be multinational, at great public expense and cost to the environment. Their hegemony is such that they have no allegiance with any particular country. Nor do they carry responsibility for the costs of environmental cleanup and restoration, or for helping respond to the disruptive wake of corporate progress—fragmented communities, unemployment, alienation, depression, violence, crime, dysfunctional families, homelessness, etc.

This is not an overstatement. The costs of corporate progress are paid for by the public, while the benefits accrue to an elite few. John B. Judis puts it this way:

> The rise of large corporations in the late 19th century led to the creation of a distant managerial elite, on the one hand, and the permanent group of wage-laborers, on the other hand. In the last three decades, the globalization of capital has removed the most basic economic questions not only from the purview of ordinary citizens, but from national governments and created (particularly in the United States) a new internationalist elite that owes its primary allegiance to global capitalism. (The new sensibility was wonderfully captured in a 1989 statement from Gilbert Williamson, president of the NCR Corporation: "I was asked the other day about United States competitiveness, and I replied that I don't think about it at all. We at NCR think of ourselves as a globally competitive company that happens to be head-quartered in the United States."[5]

The multinational or supranational corporate world and its global market economy are not viable in the long-term because they are so destructive of cultural diversity, communities, and the social economy on the one hand, and of biological diversity, natural resources, ecosystems and Nature's economy on the other. Global capitalism has established a rigid, unstable hierarchical structure which must be inverted in order to restore communities and their social economy, to prevent and alleviate present and future human suffering, to stem the irretrievable loss of indigenous and cultural diversity, and to ensure no further environmental degradation and loss of biodiversity. More laws and regulations, fines and taxes (like "polluter pays") won't work. What is needed to accomplish

these just ends is a complete inversion of the dysfunctional hierarchy of global capitalism. The necessary ideological and structural
changes that society needs to make in order to re-establish a sustainable, socially just, and equitable economy and a viable future
are as fundamental as they are profound (see Fig. 13.2). They are
relational and ethical, and are based on what I would call the
"ecopolitics" of bioethics. It is from this theoretical framework that
a cooperative global "biocracy" of interdependent communities—a
globe of villages wherein we think locally and act globally, rather
than a global village—will come to replace the dysfunctional world
of global capitalism in these regions where there are adequate human and natural resources to do so. The steps toward establishing
this alternative, post-industrial, new world "holarchy" are neither
impossible nor unrealistic. They have been clearly identified on the
basis of the criteria necessary for ensuring sustainable livelihoods
(see Addendum).

The pundits of global economism are still touting the false
hope that economic development leading to participation in the
world market is the way out of poverty and provides the job security to which all people are entitled. This hope is false because it
assumes that the very economic system that helped propel people
into unemployment and poverty will now offer to save them. David
Korten has documented with great lucidity why the public should
not trust this and should instead reclaim their political power
and reestablish localized economies. He summarizes his position
as follows:

> The global economy has become like a malignant cancer, ad-
> vancing the colonization of the planet's living spaces for the
> benefit of powerful corporations and financial institutions. It
> has turned these once useful institutions into instruments of
> a market tyranny that is destroying livelihoods, displacing
> people, and feeding on life in an insatiable quest for money.
> It forces us all to act in ways destructive of ourselves, our
> families, our communities, and nature. Human survival de-
> pends on a community-based, people-centered alternative
> beyond the failed extremist ideologies of communism and
> capitalism. This alternative is already being created through
> the initiatives of millions of people around the world who are
> taking back control of their lives and communities to create
> places where people can live and grow in balance with the
> living earth.[6]

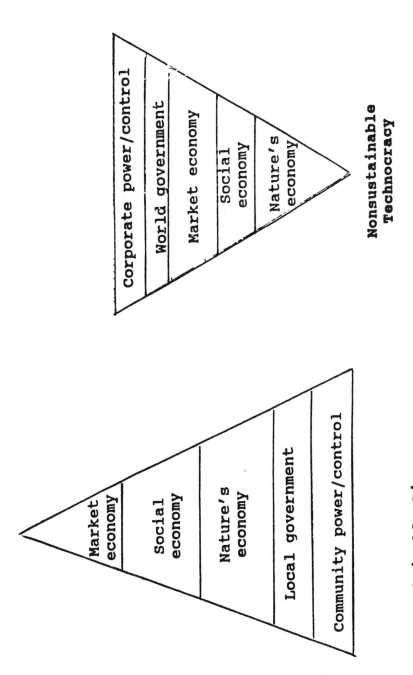

Nonsustainable Technocracy

Corporate power/control
World government
Market economy
Social economy
Nature's economy

Sustainable Biocracy

Market economy
Social economy
Nature's economy
Local government
Community power/control

Fig. 13.2. Schematic representation of economic "power pyramids" showing the inherent instability of corporate-based power and control through world government in contrast to community-based power and control through local governance of natural resources.

I prefer the term Creation- or Earth-centered rather than people-centered, so that we do not forget our responsibilities toward Nature and other sentient beings. A Creation- or Earth-centered alternative to global economism is essential, I believe, because it helps reconnect us in a covenanting and hallowing way with the life community. Our ultimate security most surely lies in this approach, where we protect Nature for Nature's sake, and respect the intrinsic value of all living things. Only that which is sacred, to quote Thomas Berry, is secure. Reverence for life, rather than for money, must be the organizing principle of society. Economic growth must become economic and ecological justice, corporate autonomy must become local autonomy, and community must include all life, because for it to be viable it must include more than the human.

Truth and Consequences

The starting point for various public policies and corporate decision makers is too often based on a false premise that profitability is the ultimate goal, and the ultimate good. Consequently, because of the way in which they think and structure reality—their worldview—the truth or validity of their position is questionable. It may be bolstered by scientific research and economic analysis, but very rarely are policies based on a decision making process that considers consequences.

A consequentialist approach entails considering past events that are linked with the present situation, either causally or correlatively, and what good or harm they caused. The Iroquois people call this "thinking seven generations back." And they speak of thinking seven generations ahead. The future consequences of any new policy or group decision must be thoroughly explored with the wisdom of hindsight. By adopting this consequentialist approach, the risks of basing far-reaching policies and decisions on a temporally isolated, historically amnesic, false premise will be minimized. For example, scientists would probably never have gathered to explore ways to utilize genetic engineering biotechnology in order to increase the productivity of farm animals so as to help feed the hungry world, had they considered past and present adverse consequences of a meat-based diet and agricultural economy. Nor would investors have been misled into financing research to develop genetically engineered animals and products in order to maintain and even increase the production and consumption of animal protein.

Unjust Government Actions and Policies

One principle of democracy is that government and its leadership of publicly elected public servants should serve the public good. The erosion and virtual annihilation of this principle is nowhere better demonstrated than in the response of government to the concerns and legislative initiatives of the animal and environmental protection movements. It would seem that the public good is either being ignored or recast to serve primarily private interests. The public's rights to a clean environment, to the preservation of natural ecosystems, to food, air, and water uncontaminated with veterinary and agrichemical drugs and industrial wastes, and to products (such as cosmetics and household cleaners) that have not been tested on animals are being violated. While government is hard-pressed to demonstrate scientifically and ethically that the public good is best served by ignoring these concerns, the final denominator is economics. The argument that the economic benefits to the public—more jobs, cheap food, "safe" consumables—override the costs (to the environment, to animals, and to the public's ultimate health and well-being) has become patently absurd.

This absurdity is exemplified by some recent instances of government pandering to private interest groups (and their Political Action Committees and well-heeled lobbyists):

- Refusal to enforce the Marine Mammal Protection Act and set up trade barriers against Japan and other countries that still engage in whaling and drift-net fishing;

- Refusal to accept international standards on global warming agreed upon by all other industrial nations to help reduce carbon dioxide emissions because these would harm the U.S. economy;

- Refusal to prohibit the export of pesticides to the third world that are so hazardous that their use in the U.S. is banned;

- Refusal to initiate legislation to prohibit the addition of antibiotics to farm animals' feed which, for consumer health reasons, has been enforced in the European Economic Community for several years.

These instances illustrate clearly that in the U.S. the elected government serves the interests of the private sector that collectively represents an anti-democratic technocracy. Government inertia and indifference has led the more militant animal and environmental protection groups, like the Animal Liberation Front and Earth First!, to acts of violent civil disobedience. Anarchy is the product of a totalitarian technocracy that places private economic interest over the public interest. That non-profit public interest groups have to spend their limited financial resources to sue state and federal agencies to enforce animal and environmental protection laws is evidence enough that the government does not serve the public good. These public interest groups are further limited by their non-profit, tax-exempt status from engaging in lobbying activities to further legislative reforms and initiatives.

While it is the role of government agencies to enforce the law and serve the public good, it is not their role to determine public policy without any directive from Congress. Yet the U.S. Patent and Trademark Office recently made the unilateral, arbitrary, and capricious determination that all genetically-engineered life forms, including the entire animal kingdom, can be patented. This determination, in the total absence of any public debate, Congressional hearings or directive from Congress, is nothing more than a concession to the biotechnology industry. The State Department scuttled Congressional debate on this issue and attempts to legislate a moratorium on animal patenting that would have allowed time for more discussion on its ethical and socio-economic consequences. The State Department insisted that the patenting of animals was essential to protect the U.S. economy and to give the U.S. a competitive edge in a highly competitive world market for new products and services. In other words, to oppose the patenting of animals was against the national interest, if not the nation's security, and was therefore unpatriotic.

Even before it was approved for use by the Food and Drug Administration, the U.S. Department of Agriculture's (USDA) Secretary was urging the European Economic Community to accept rBGH. Yet this drug's safety and its socio-economic impact had not been fully determined. The USDA's responsibility for enforcing the Animal Welfare Act has a long history of non-enforcement and complicity with private interests. USDA's rare intervention in the commercial puppy mill industry, where purebred dogs are mass produced under all too frequently filthy and impoverished conditions, reflects the prevailing posture of the government to not interfere in matters of commerce. If the ends are profitable, provide employment, and increase the public demand for goods and services (in the case of puppy mill dogs, this means pet foods and veterinary medicines), then the means are almost invariably accepted without question.

The USDA panders to other interests such as cattle ranchers, as well, leasing millions of acres of land held in public trust at cut-rate prices and with little regard to the devastating effects of overstocking and overgrazing. This government agency actually aids and abets the degradation of public lands and loss of biodiversity through its little publicized Animal Damage Control division. This division in concert with state agricultural and wildlife agencies has virtually exterminated the grizzly bear, wolf, and mountain lion from the North American continent, and continue to harm many wildlife species in its costly and ineffectual quest to eliminate the coyote from public land with traps, poison baits, aerial shooting, and den bombing. This mirrors the Department of the Interior's policy of allowing ranching, hunting, and trapping on lands designated by Congressional legislation as National Wildlife Sanctuaries.

It would seem that any financial claim or interest based upon the commercial exploitation and commoditization of animals, the environment, and any natural resource takes precedence over all else, in the eye of the government. Rather than living by the ancient wisdom of the Golden Rule, the new rule is that those who have the gold rule. Governmental complicity with private industry is underscored by the enormous funds the latter provides to ensure that key positions in the government (that are ostensibly determined by public election of candidates for office) are occupied by their chosen representatives. That these representatives have close ties as employees, consultants, and board members of private corporations, prior to election and after their term of office has expired, is also a matter of public record.

Complicity with private industry to further corporate interests without regard for the will and consensus of public opinion is evident at all levels of government, from the federal and state levels to the municipal and the judicial. The Supreme Court has ruled that people and animal protection organizations have no legal standing to sue on behalf of animals if they are not harmed financially or in other direct ways, for example, by a corporation that poisons and kills animals to "safety" test some new household cleaner or cosmetic.

When we look at the current status quo of animal and environmental exploitation today, it is clear that the means whereby private interests are secured and satisfied are discounted if the ends are profitable. This attitude is evident in the highest circles of government. Witness the statement of a 1983 presidential task force that "health, safety and environmental regulations should address ends rather than means."[7]

Defense of this status quo is reaching a new level of intensity as articles in *Readers' Digest* and other periodicals unconditionally tout the belief that medical progress will be arrested if the animal protection movement achieves any legislative success in limiting the numbers of animals used in biomedical research. The American Chemical Manufacturer's Association, the lobbying arm of the petrochemical-pharmaceutical industrial complex, had a budget of $15 million for 1990 to spend on ensuring that no legislation passed that would limit the use of pesticides. Agribusiness farm journals tell livestock and poultry producers to join them in their battle to discredit the animal protection movement and block any farm animal welfare legislation. They use as their argument the claim that this movement is the reason why animal agriculture has become less and less profitable for farmers and ranchers, hundreds of thousands of whom have gone bankrupt over the past decade. The animal protection movement is essentially being used as a scape-goat to smokescreen the real reasons why family farms and rural communities are failing across the nation. The primary reason is that the family farmers' interests are irrelevant to the interest of agribusiness and the allied petrochemical-pharmaceutical industrial complex, which essentially controls government and public policy.

Professional organizations like the American Medical Association and American Veterinary Medical Association (AVMA), which are part of this technocracy, are committed to protecting the status quo of animal exploitation by agribusiness and the biomedical industry. At the 1990 annual conference of the AVMA, which has regularly included the Association of Veterinarians for Animal Rights (AVAR), members and participants in AVAR found that their hotel rooms and meeting room had been canceled, that the announcement of their meeting in the AVMA's conference program had been deleted, and that a conflicting "bioethics" symposium had been scheduled for AVAR's old time slot.

Animal scientists, veterinarians, and other "experts" from academia and various government agencies have repeatedly presented testimony to block any legislation that might change the status quo of animal exploitation. In congressional hearings over legislation to enlarge the crates of calves raised for veal to allow the calves at least to turn around, lie down, and get up easily, one animal scientist testified that "there is no scientific evidence that the behavioral needs of veal calves are not satisfied under existing conditions." A California state public health official, responding to public concern over spraying residential areas with Malathion to control fruit flies, echoed this obfuscating rhetoric, saying that "there is no scientific evidence that any pesticide has caused cancer in human beings, and the State's agricultural interests must be protected."

That "health, safety and environmental regulations should address ends rather than means" is all very well on paper. But when the public cries "foul" over the means and the government remains indifferent, saying it is in the best economic interests of the state or nation, or in the best interests of the public not to question the means whereby certain ends are achieved, public officials should be censored and held accountable. The ancient proverb, "No just ends can come from evil means" seems alien in these times. The promise of a "kinder, gentler nation" rings hollow indeed to those who care for creatures and Creation and see no virtue in harming the environment and fellow creatures for short-term financial gain.

From Violent Times to Gentler Ways

We live in violent times. We are neither immune from the consequences of violence, nor are we wholly innocent. As citizens of a violent society—a society whose industrial economy unjustly and unjustifiably violates the sanctity of life and the integrity of Creation—we are participants in a violence and destruction that is intensifying and spreading globally. We may feel genuine despair and hopelessness, believing that as individuals we are helpless in the face of violence and injustice on every front, and that to try to make the world a better place is naïve idealism.

To make no attempt to make a difference is unconscionable denial and irresponsible selfishness. No matter what our station in life may be, no matter how rich or poor, old or young, oppressed or free, there are no grounds for inaction and no basis for despair and hopelessness. We *can* help ourselves and each other reduce the violence and destruction in our society by changing our own life-styles, consumer habits, and all manner of activities and attitudes that contribute, directly or indirectly, to the perpetuation of violence.

It is amazing how surprised people are when they are victims of violence, and disgusted and outraged when they experience it vicariously via the daily news. Yet they seem to be completely unaware of how they are part of the tangled web of the cause and consequence of violence. This web, into which we are all born, has persisted for generations. It is as tenacious as it is almost invisible. But as we begin to see it, to understand how we are connected directly and indirectly with so much of the violence and injustice in the world today, it becomes less tenacious. It begins to lose its hold on us once we acknowledge that we are part of the web and that we are powerless if we use force to try to free ourselves. We cannot stop violence with violence. The metaphor of the fly becoming more tangled in the spider's web the more it struggles is perhaps a lesson from Nature that affirms the wisdom of the Sermon on the Mount where Jesus of Nazareth urged the people to "resist not evil."[8] Indeed realizing the alternative path and higher power of compassion, justice, and nonviolence entails an almost effortless stepping out from this metaphorical tangled web of ancient and continuing human violence and injustice against humanity, creatures, and Creation.

To begin to live more gently so that others may simply live necessitates being informed, if not educated, and concerned, if not ascetic. Why should one begin the process of renouncing violence and injustice, and by implication, renouncing a violent consumer life style? Because it is enlightened self-interest to do so, and because it is surely a matter of personal integrity and dignity to be humane. The alternative, to be inhumane, is to be subhuman. For urban consumers to adopt a more or less organic, vegetarian diet where animal fat and protein are not dietary staples, for example, will greatly improve public health and reduce the burden of health care costs to society, improve the well-being of farm animals, and in the process support those producers who want to farm without harm, and to help restore and protect the environment.

It is defeatist to condone forms of human violence and injustice, especially those against minorities, animals, and Nature, often falsely justified on economic grounds. People justify violence by reasoning that since violence occurs in Nature, it is a natural and unavoidable aspect of human life. Big fish eat little fish; Nature is red in tooth and claw. So we come to model our own ethics on the amoral food chain "eco-logic," placing ourselves at the top and legitimizing much violence and injustice on the grounds of human superiority, entitlement, progress, and necessity. This path of violence and rationalization of the dominant way is ultimately life-negating.

Following the path of nonviolence and the gentle way is life-affirming. A life-affirming step is to make a detailed, personal moral inventory and begin to make amends by changing those actions, attitudes, needs, and wants that cause harm to ourselves, to others (including animals), and to the environment. We all have the power to choose to use compassion and justice as the cardinal points on our ethical compass, regardless of our status in society, and in what culture we may be living. Justice is one of the universal and universalizing principles of bioethics. The challenges and opportunities in promoting and adopting global bioethics worldwide are the subject of the next and last chapter.

ADDENDUM

The following principles and public policies necessary for establishing a humane sustainable society that I helped draft at the Sustainable Livelihoods Consultation, North America, January 13–15, 1995, in Washington, D.C., before the 1995 International Development Conference.

Principles of Sustainable Livelihoods

In an era of global social crisis characterized by increasing unemployment, jobless growth and ecological destruction, we need a broader vision of how people can meet their needs in a sustainable way. Attempting to solve the world's employment crisis using conventional job creation through sustained economic growth cannot work.

The concept of *livelihood* defined as "a means of living or of supporting life and meeting individual and community needs," provides new perspectives on developing healthy sustainable societies that provide people with secure and satisfying livelihoods. Sustainable livelihoods are based on a web of functional interrelationships in which every member of the system is needed and participates. Sustainable livelihoods provide meaningful work that fulfills the social, economic, cultural and spiritual needs of all members of a community—human, non-human, present and future—and safeguards cultural and biological diversity. The following is not an exhaustive listing of the components of sustainable livelihoods but an attempt to identify the key determinants.

Sustainable Livelihoods:

Promote equity between and among generations, races, genders, and ethnic groups; in the access to and distribution of wealth and resources; in the sharing of productive and reproductive roles; and the transfer of knowledge and skills.

Nurture a sense of place and connection to the local community, and adapt to and restore regional ecosystems.

Stimulate local investment in the community and help to retain capital within the local economy.

Base production on renewable energy and on regenerating local resource endowments while reducing intensity of energy use, eliminating overconsumption of local and global resources and assuring no net loss of biodiversity.

Utilize appropriate technology that is ecologically fitting, socially just and humane, and that enhances rather than displaces community knowledge and skills.

Reduce as much as possible travel to workplace and the distance between producers and users.

Generate social as well as economic returns, and value non-monetized as well as paid work.

Provide secure access to opportunity and meaningful activity in community life.

These principles encompass a holistic set of values that are non-exploitative and promote participation in decision-making; emphasize the quality and creative nature of work; place needs over wants; and foster healthy, mutually beneficial relationships among people and between people and their environment (especially domesticated animals). It is hoped that these principles and their underlying values can stimulate further discussion.

Public Policy

Sustainable livelihoods are supported by political, economic and social policies that enable mutually beneficial relationships to develop among people and the whole community of life. Economic globalization, on the other hand, primarily advances supranational corporate interest, and is often inimical to human and environmental well-being. Current policies externalize social and environmental costs, destroy ecosystems, pit localities into competition with one another, and lower standards. Current measures ignore many of the crucial social functions on which all economies depend, in particular women's tremendous productive and reproductive roles. Policies are now geared toward economic growth based on overconsumption by the few while the needs of the many go unmet. Instead, socio-economic security and equity, meeting the *needs* of all and promoting authentic human development should be the overall goals of policy formulation.

Policy formulation should begin with visioning processes that involve all sectors of community, as decisions made by all stakeholders better ensure equity, human rights and effective implementation. Central to a broad policy framework that supports sustainable livelihoods are:

- an investment in people and the environment as well as in physical capital;
- explicit recognition that women's empowerment is central to the achievement of broad-based socio-economic goals;
- broad public participation in the establishment of research priorities and the assessment and selection of technologies consistent with needs of sustainable communities; and
- new resource accounting and institutional mechanisms for resource allocation and debt management and relief.

Political Priorities

Sustainable livelihoods require public participation and involvement in policy making at all levels to keep government agencies and officials responsive and accountable for their decisions and actions. Political reforms should both limit and make transparent the influence of corporate lobbies and campaign contributions. Corporations should be held accountable to a code of conduct based on principles of social and environmental responsibility. Multilateral trade agreements, treaties, and conventions should not supersede local, state, and national sovereignty. Subsidiarity should be an organizing principle of government, supporting the local rootedness of livelihoods.

Economic Priorities

To promote sustainable livelihoods, power must be rooted in the localized economies. Economic policy should be based on full-cost accounting which incorporates social and environmental costs and benefits. Trade agreements and tax pollicies should favor local needs over export marketing, encourage sustainable production and consumption, and support renewable resource technologies. Such policies will support worker rights, debt relief, and local control over resources within a framework of broader responsibility to share and protect resources.

Socio-cultural Aspects

Socio-cultural policies should support principles of sustainable livelihoods in education, health, arts and the media, drawing on the wealth of cultural diversity and encouraging exchange of indigenous and modern knowledge, wisdom and skills. Special attention must be given to transforming structures that perpetuate inequity, injustice and intolerance, including those that perpetuate inequality and injustice toward women.

CHAPTER 14

Toward a Universal Bioethics:
Challenges and Opportunities

The bioethics of an ecological democracy should never be compromised. These can be the keystone for a new, egalitarian economic paradigm of sustainable development. There surely are moral and ethical absolutes and principles that are binding and universalizing, rather than authoritarian and anti-democratic, like respect for justice, human rights, animals and the environment.

Can compassion and reverential respect for life ever be compromised? Are these not the integrating universal ethics of all people? According to a recent cross-cultural study by bioethicist Darryl Macer, such ethical unity is not only possible, it already exists to a considerable degree. This unity I would call a pluralistic monism, because from a plurality of cultural and religious traditions and perspectives, a shared reverential and compassionate respect for all life is clearly evident. Macer found similar concern and ethical sensibility regarding biotechnology, environmental destruction and cruel animal exploitation in Japan, Israel, Australia, and many other countries.[1]

The basic bioethical principles of a humane and sustainable society and a new world order cannot be compromised as they are today by the governments of the industrial world that claim to be committed to helping this new world order come to fruition. Both good science and ethics are being compromised, often with the best intentions, by those who promote the ideology that economic growth and industrial expansion are synonymous with conservation and with the well-being and security of the human race. The ideology of technocratic imperialists becomes paternalistic when first confronted, then becomes racist, sexist, and speciesist when opposed. Like the materialistic and hierarchical philosophy of Aristotle that

205

Vice President Al Gore embraces in his book *Earth in the Balance*, this ideology is crassly materialistic, economically deterministic, and enchanted with technology.[2] It appears that the Emperor's new clothes and waving flags for environmentalism, minority rights, and endangered species are as transparent as ever.

Even with the best intentions, no administration can make this world a better place if the will of the people is not with it. And what is the will of the people? Countless polls have been conducted and analyzed by sociologists, economists, and others for decades. Synthesized by Abraham Maslow, they show that when basic needs are satisfied, the human species seeks a self-actualizing path that is ethical and spiritual rather than egotistical and materialistic.[3] This more mature stage of human development has yet to be manifest at the community and corporate level in contemporary industrial society, because as Maslow showed, it is thwarted by a host of forces that are antithetical to our self-actualization and thus in opposition to the natural ethos and will of the people.

These forces are evident in the Clinton administration's approval in 1995 of Norway's continuing to slaughter whales in the name of sustainable economic growth and with the belief that such killing is based on good science. This approval is contrary to the will of the people, a will and consensus that is reflected in the publicly crafted and supported Marine Mammal Protection Act. The heartless rhetoric and rationalizations of science-based policies and newspeak slogans like "win-win resolutions" seem like propagandistic lies. But they are, in fact, the value-based constructs that make up the reality and truth many people live by. Many pervert science and ethics to serve their own pecuniary interests and justify further destruction of the Everglades, the old growth forests, and the endangered animal, plant, and human communities around the world that are variously seen as wild, undeveloped, and uncivilized.[4] To these insults to our intelligence, we must add the new foreign aid programs for sustainable development, and international aid ostensibly aimed at alleviating poverty. These pay no regard to means and ethical costs and consequences, but are primarily intended to sate the nonsustainable appetites of multinational corporate interests and to promote the ideology of economic development and industrial expansion—measures that the earth cannot sustain.[5]

The greatest poverty is not material but spiritual. It includes despair, ethical blindness, and disintegration of the moral fabric of society. Antisocial and sociopathic behaviors are linked with a lack of conscience. They are distinct, though, from the alienation of those who are not ethically blind and not afflicted by the spiritual inertia of a morally bankrupt society. In a purely materialistic society where economic determinism has no regard for ethics and makes the marketplace the final arbiter of all values, alienation and sociopathic behavior are somewhat understandable reactions.

Technological developments and industrial expansion now take place virtually in an ethical vacuum. The quasi-scientific principle of amoral objectivity is embraced to legitimize as scientific the belief that the ultimate good of society will be secured by industrial expansion and the promise of ever more jobs, products, and services. But the elusive good life that this worldview promises is like a mirage in the desert. It is an illusion of progress that encourages consumption and the spread of deserts and bioindustrialized wastelands that cannot sustain the hungry multitudes in body or spirit.

Moral Absolutes

Aware of these developments and their serious ethical and moral causalogy, Pope John Paul II released an encyclical in 1994, *Veritatis Splendor* (*The Splendor of Truth*), in which he correctly insists that an individual's conscience cannot alone determine which actions are moral, since so many disturbed people today seem to have no understanding of good and evil or remorse. So he proposes a number of absolute moral truths to guide our behavior. He lists the evils to be avoided, and they include torture, slavery, and genocide. Unfortunately, he also lists avoiding the use of contraceptives.

The environment and animal and plant kingdoms are seriously threatened today by human overpopulation and overconsumption.[6] Taking care of the Earth in part by using contraceptives is surely not a sin, since the harm, or evil, that arises in the absence of any family planning is one of the greatest tragedies of our species. Is it not immoral to propose otherwise? To discount contraception on the grounds of faulty situational ethics that are subordinate to the moral absolutes of *Veritatis Splendor* is destructive. Appropriate and compassionate ethical choices in times of crisis and in difficult situations are not automatically situational ethics. Rather, as in the case of euthanasia and contraception, such decisions are consonant with trans-situational, transtemporal, and moral absolutes, namely, ahimsa and reverential respect for Earth's Creation. The moralistic absolutism of condemning contraception carte-blanche undermines the role that these important moral absolutes could play as the ethical foundations of a truly civilized society.

There is an operational distinction between ethics and morality that has been well expressed by Darryl Macer and co-authors in the book *Bioethics for the People by the People*:

> **Ethics and Morality**. Both words mean similar things (based on words for "custom" in Greek and Latin respectively), but moral philosophers use them in different ways. Ethics is used to refer to the critical study of morals or morality, the latter being the specific values and behaviour of individuals or groups. Ethics itself does not promote a particular viewpoint, but is concerned with looking at the assumptions behind differing moral choices, seeking to clarify the arguments and the concepts that are used when people justify their moral views.[7]

As Socrates observed, "The unexamined life is not worth living." Indeed, if the purely materialistic, instrumental, and self-serving values and ethos of industrial consumer society remain unexamined by its technocratic priesthood and proletariat, then life will indeed not be worth living. A shallowness of values and dearth of ethics demean life and Creation. And by evaluating things as either oriented towards human ends or as useless, human life likewise becomes demeaned as expendable if it is not useful.

The technocracy fails to realize that what they see as a failure of morality and of family and community values is symptomatic of a much deeper ethical and spiritual crisis. This crisis is a product of the state of mind that gave rise to technocratic imperialism. It clearly cannot be rectified by that same state of consciousness, but can only perpetuate with new palliatives and social programs. Macer emphasizes:

> We are currently in a crisis of domination, not just an ecological crisis, but a crisis of our whole life system, brought upon the entire globe by ourselves. The origin of this crisis is in human behavior and attitudes, and the tremendous power of our technologies to shape the world. As a reaction against this, some people attack what they see as the cause, science and technology, and its effect upon people's philosophy; however, the real cause is the age old problem of human

selfishness, which has become embedded in the short-term economic desires of many businesses and governments.[8]

Ethics and Trade

A quasi-ethical framework can be fabricated on primarily economic criteria, under the banner of "sustainability." This is what has happened with GATT and the WTO, and much of the international accord concocted by the 1992 United Nations' Conference on Environment and Development, otherwise known as the Rio Earth Summit. From the narrow materialistic perspective of the GATT participants (who subsequently under pressure from public interest groups have promised side-agreement correctives that address social and environmental concerns), a new world order for the human species was completed and ready to fly under the flag of world free trade.

As the final round of GATT negotiations came to an end, many Third World countries vehemently protested attempts by industrialized democracies to impose trade rules to protect worker rights, set limits on child and prison labor, and promote collective bargaining rights and pollution controls. These steps were dismissed as protectionist, designed to "rob poorer nations of natural competitive advantages that offer the basis for future growth," according to reporter William Drozdiak.[9] These disturbing reactions by a block of fifteen developing nations from Africa, Asia, and Latin America underscore the difficulties in trying to ethicize purely materialistic values and goals within a competitive, market-driven world economy. One might at least expect some unanimity over child labor and pollution control, but not when one realizes that situational ethics vary according to the degree of poverty and human need. Clearly a closer examination is warranted of the kind of life that these developments in world trade and a global economy will create for us and for future generations. Examining motives, means, and ends is a start, for it enables the identification of areas where a bioethical framework is either deficient or entirely lacking.

Motives, Means and Ends

Is it better to treat an animal humanely for its own sake or to preserve one's sense of humanity? Is it better to protect the environment for Nature's sake, or for reasons of personal health or economy?

It is important to evaluate our motives since they can either converge on a mutually beneficial end, or lead us down the slippery
slope of increasing selfishness and related conflicts of interest, like
justifying some forms of animal cruelty or environmental damage in
the name of medical or agricultural progress or unavoidable necessity. It is important to understand that we often have more than one
reason or motive that influences our choice of action and consequence. Ethical decisions are not always pure—more than one single
motive often influences our choice. Through impartially evaluating
our own personal inventory of motives, or motivational states and
associated values, we can avoid inner conflict and achieve some degree of ethical consistency and confluence in our lives. We treat animals humanely for their sakes and for ours just as we protect the
environment for our sakes as well as for Nature's because it is enlightened self-interest to do so.

But the task of conflict resolution and goal of priority setting
gets more difficult when dealing with others who have very different motives or reasons for wanting the same or different ends. If
the same end or goal is desired and shared by people who do not
share the same motives or values as you and I, then as moral
purists, we might not want to have anything further to do with
them. But as ethical pragmatists, we might accept some motives
that accomplish the ends that we desire. As a moral purist, one
might oppose contraception, mass abortion, and gerontocide as unethical responses to overpopulation, but as an ethical pragmatist
one might accept contraception and euthanasia. These are examples of various means that not all people accept, used to achieve an
end that most do accept. This difference in means is reflective of
different values. Differences in means can also reflect different motivations. For example, hunters and loggers could join with conservationists to save wetlands and forests, but their motives may be
quite different. A moral purist might scorn coalitions of such unlikely bedfellows as hunters, loggers, and conservationists. But the
ethical pragmatist first asks, what is the best and most realistic
way to save the birds and the forests, and what other viable alternatives are there?

The pragmatism of bioethics allows for considerable accommodation of others' views and motives when a common goal is agreed
upon. But we should not assume, therefore, that others who desire
the same end as we do so for the same reasons. Accommodation
and agreement is facilitated when there is honesty and respect for

the right of others to hold different views. Then the ground may be laid for the democratic, multilateral adoption of basic bioethical principles.

This democratizing or egalitarian openness to others' motives, means, and ends, demands humility by those who believe that they are the most righteous, and self-control by those, often in the minority, who have the most power. As world trade and governance evolve through the GATT and NAFTA, these essential principles of global accord and sustainability, humility and self-control, need to be acknowledged and practiced. The behavior, motives, means, and ends of governments and transnational corporations, like those of tribal peoples, reflect a consensus that can be challenged and changed by personal example.

In his life and works, Mohandas Gandhi sought to bring ethics into the realms of politics, industry and finance. His views are as relevant today as they were years ago, especially considering the neocolonial nature of GATT and the WTO. In *My Socialism* he states:

> I must confess that I do not draw a sharp or any distinction between economics and ethics. Economics that hurt the moral well-being of an individual or a nation are immoral and, therefore, sinful . . . True economics . . . stands for social justice, it promotes the good of all equally including the weakest, and is indispensable for decent life.[10]

Gandhi's principle of *swadeshi* is especially germane to discussions of protectionism and sustainable development. This principle of self-sufficiency emphasizes the frugal use of local resources rather than building an economy that relies on the input of outside raw materials and other resources. This principle is exemplified by what contemporary agriculturalists call sustainable agriculture. Gandhi contended:

> Swadeshi is that spirit in us which promotes the use and service of our immediate surroundings to the exclusion of the more remote . . . In the domain of politics, I should make use of the indigenous institutions and serve them by curing them of their proved defects. In that of economics, I should use only things that are produced by my immediate neighbours and serve those industries by making them efficient and complete where they might be found wanting.[11]

Gandhi foresaw a network of interdependent but largely self-sufficient village communities, a Jeffersonian vision indeed. He wrote:

> My idea of village swaraj is that it is a complete republic, independent of its neighbors for its own vital wants, and yet inter-dependent for many others in which dependence is a necessity. Thus, every village's first concern will be to grow its own food crops and cotton for its clothes . . . My economic creed is a complete taboo in respect to all foreign commodities, whose importation is likely to prove harmful to our indigenous interests. This means that we may not in any circumstances import a commodity that can be adequately supplied from our country.[12]

Like Gandhi, E. F. Schumacher, very much influenced by the compassionate and non-violence doctrines of Buddhism, endeavored to weave these principles into economic theory and practice. He contends that "non-renewable goods must be used only if they are indispensable, and then only with the greatest care and the most meticulous concern for conservation. To use them heedlessly or extravagantly is an act of violence, and while complete non-violence may not be attainable on this earth, there is nonetheless an ineluctable duty on man to aim at the ideal of non-violence in all he does."[13]

Books like *The Case Against "Free Trade": GATT, NAFTA and the Globalization of Corporate Power* by Ralph Nader, et al., clearly identify the lack of any bioethical framework to maximize the good and minimize the harm of the evolving new world order.[14] As Wendell Berry states in this book, "The proposed GATT revisions offend against democracy and freedom, against people's natural concern for bodily and ecological health, and against the very possibility of a sustainable food supply."[15]

The motives, means and ends of those individuals, corporate bodies, and agencies involved in the Rio Earth Summit, GATT, and the WTO are indeed more lamentable than laudable. They are lamentable not so much in terms of evil intent but of evil consequence by virtue of an arrogant and righteous belief that their way is best, and that whatever harm may be done is a small price to pay for material progress, industrial expansion, and economic growth. But in the final analysis, more than money matters. The very dignity of the human species, so easily perverted by materialism and enchanted by technology, is preserved by our singular and collective opposition to the evil that men do, often with the best intentions, but interminably to each other, animals, and Nature. We may find some encouragement in the conclusion of Macer that "universal bioethics already exists at the level of individual decision-making, and therefore it is certainly possible to develop social and educational systems to allow universal ethics at the higher level of social systems."[16] The evil that arises from industrialism's technological imperialism and economic determinism may be a blessing in disguise, by unifying humanity against those harmful elements in our own nature and towards making the next evolutionary step toward a more humane and significant way of life.

Science Is Truth: The Newspeak of Animal and Environmental Exploitation

In his prophetic book *1984*, George Orwell described a process that he called "double-think," which is essentially a twisted perversion of reason to serve the interests of those in power. Through this process new words and slogans were contrived, like "war is peace." He called this contrivance "newspeak," which was used to sanctify unethical policies and activities.[17]

Just as Orwell foresaw, new words and slogans are used by the technocracy to make unethical activities acceptable to the public. Today we have the newspeak slogan "science is truth," and anyone daring to question it is dismissed as an anti-science, anti-progress neo-Luddite. The double-think that leads a person to believe that science is truth has become widely accepted by academia, professional organizations, and government. The scientism of relying on quantifiable numerical indices is held to be objective and impartial. For example, the American Veterinary Medical Association (AVMA) has established the policy that animal welfare is a concern that only science can address. Animal welfare is seen not as an ethical issue, since ethics are subjective, and not impartial. Rather, the AVMA holds that animal welfare should be determined by quantifiable and measurable scientific indices.

The Food and Drug Administration followed the science-based approach of advocates of the genetically-engineered rBGH in approving its use. Since science is truth, nonquantified and subjective considerations like concern for the welfare of cows and the socio-economic impact on small dairy operations, are dismissed as unscientific and irrelevant. The sole responsibilities of government are to determine product effectiveness and consumer safety.

Yet even consumer safety, as with the approval of chemical pesticides, can be discounted when a science-based, risk-benefit analysis is done. The probable risk of a few people developing cancer, or infants being born with birth defects, are outweighed by the societal benefits of a plentiful supply of food, according to the double-think logic that precludes ethical concerns. Language can be changed not only to justify unethical behavior but also to make certain activities appear more benign. It is no coincidence that the USDA is dropping the word "slaughter" and using the term "harvest" when speaking of livestock and poultry "pre- and post harvest" inspections.

The exclusion of bioethics from the worldview "science is truth" is disturbing on many counts. It is reflective of an amoral society and technocratic power elite that misuses the scientific method. Such misuse, which I call scientism, is linked with the arrogance of humanism and economic determinism, to justify the adoption of policies and practices that may be unethical.[18] For example, Vice President Al Gore told the press that Norway's move to have the whale moratorium lifted to permit the "harvesting" of minke whales in 1994 was based on "good and sound science."[19] This policy position, which had been developing under pressure from Japan, Norway, and Iceland (countries that already have permission to kill whales for scientific purposes), is justified on the basis of yet another newspeak slogan: sustainable development.

An Antarctic sanctuary for whales, where whaling would be prohibited, was proposed at the International Whaling Commission meeting in Mexico. According to one news report, "The sanctuary plan was fiercely opposed by Japan, which—at present nominally restricted to *research* whaling—wants to resume limited commercial whaling. 'Whale stocks are protected whether they are healthy or not,' complained one Japanese delegate. 'This is not scientific.'"[20] The important concept of sustainable development becomes perverted by self-interest when ethical concerns are excluded from consideration—concerns like the ethics of killing whales when their numbers indicate that they are no longer endangered.

On the whale issue, and on the broader issue of protecting rare and endangered species and ecosystems, we find evidence of double-think. For instance, it is reasoned that with good management, oversight and quotas, allowing the sustainable harvesting of whales, and elephants for their ivory, is the best way to protect the species. That wildlife must pay its own way is the kind of economism that is gaining acceptance under the twisted double-think. It is rhetoric in the name of conservation.

The ideal of sustainable development may never be achieved if simplistic economic and scientific formulations and analyses are used as the criteria for progress.[21] Biases caused by pecuniary interests and ethically impoverished paradigms or worldviews, supported by contrived scientific validation, are cause for concern. Until science and ethics are given equal value and consideration, it is very unlikely that national and international wildlife conservation, and environmental and animal protection efforts, will ever make a significant difference.

The least appropriate ethical and scientific response to these issues is the simplistic approach. It is insufficient to ask if the killing and exploitation of animals, wild and domestic, is either right or wrong, or sustainable or not. A broader, bioethical perspective is needed, beginning with the basic question of whether it is necessary. What alternatives are there to killing animals and subjecting them to suffering in the process of exploitation? A second important question to ask, one which goes beyond reductionistic environmental impact assessments, is to determine whether such actions contribute to protecting and enhancing ecological integrity and biodiversity. This form of ethical enquiry challenges and demands scientific documentation and validation, leaving the onus of responsibility and proof firmly in the court of those who seek to profit from exploiting animals, and not the defenders of animals' rights and interests.

As Darryl Macer has written, "Bioethical decision-making involves recognition of the autonomy of all individuals to make free and informed decisions providing that their decisions do not prevent others from making such decisions. This tenet is consistent with democratic principles, and the extent to which a society has accepted it is one criterion of the success of bioethics.[22] Clearly, the success of bioethics is contingent upon full and informed public involvement in the decision-making process on a host of issues, from whaling to genetic engineering. As we move toward a more globally integrated economy, via GATT and the World Trade Organization, the establishment of an international bioethics forum, as Macer urges, is needed to enhance global sustainability and to protect biocultural diversity. While it is true that the U.N. Declaration of Human Rights recognizes some universal, cross-cultural bioethical principles, these must now be extended and established for non-humans and the environment.

Respecting Others' Views

Theologian Chris Chapple makes the following relevant point: "Quite often it is assumed that ethics proceeds from philosophy or theology, that the superiority of a philosophical or theological system results in a superior system of ethics. However, given the shared biology common to the human condition, it stands to reason that a shared ethic would be more palatable to diverse cultures than a shared ideology."[23] Ideological diversity need not be sacrificed, or be a barrier to conflict resolution, or to cooperative problem solving when an ethic such as reverential respect for life or ahimsa is shared. Chapple emphasizes that the Jaina principle of anekantavada, respect for the opinions of others, is the essence of intellectual non-violence. We can agree to disagree and from that point people with opposing views can begin to collaborate on finding a constructive resolution. Philosopher Stephen Clark wrote to me saying, "The best way to live (which we call divine) is *not* to force others to live as we think best, but to accommodate their choices, even at cost to ourselves (see *Philippians* 2: 6–8)." But this can be problematic when others choose to treat their own kind or other animals inhumanely. There are *no* 'two sides' to another being's suffering. *Suffering is suffering*, and when it is caused deliberately for reasons of custom, profit or whatever, to accommodate such reasons and do nothing is morally reprehensible.

An important component of openness to another's views, and to conflict resolution, is to avoid labeling others in terms of their ideological, racial, occupational, or other affiliations. Labeling can be highly polemicizing and can bring with it the pejorative baggage of negative expectations and associations. Labels are often used to fuel public fears and ridicule: Humanitarians are called "bambi-syndrome humaniacs," and environmental protectionists "tree huggers," both affiliations misjudged as favoring animals and Nature over people and jobs. Those who question technological materialism, rational humanism, and scientific imperialism are dismissed as anti-social neo-Luddites opposed to progress and ignorant of the veracity of science and benefits of economic growth and industrialism. Demeaning labels that obfuscate the wisdom of others' values, motives, and worldview is the way of fundamentalism and totalitarianism, which in the final analysis is anti-democratic and counter-progressive. Creative accommodation of the plurality of human values and belief systems is as vital to individual freedom and fulfillment and to the functional integrity of society as, by analogy, biodiversity is vital to the functional integrity of ecosystems. The loss of this plurality leads to totalitarianism and a

dysfunctional society, just as the loss of biodiversity leads to the ultimate collapse of natural and agricultural ecosystems.[24]

The worldview of technocratic materialism excludes moral and
ethical considerations, and discounts subjective emotional and esthetic concerns. Likewise, the utilitarianism of instrumental rationalism is purportedly scientific, but in actuality is amoral and
objective. It may even discount established laws, statutes, conventions, and historical precedents, arguing that situational rather
than categorical ethics are the only reliable criteria to determine ob-
jectively the acceptability of various activities, products, processes,
and policies. The technocratic view that science knows best flies in
the face of public distrust and opposition to such activities as the creation of organ donor pigs and genetically engineered and irradiated
food. The public's distrust is based less on scientific ignorance than
it is on a historical memory of the harms that many new technologies and products, like hydroelectric dams, pesticides, irradiation,
and chlorofluorocarbons have caused in the past. The "forward
thinking" technocrats unfortunately suffer from historical amnesia
of these issues. They deny that their motives and worldview stem
from the same historical roots that the public distrusts.

Technocratic "objectivity" cannot contain or comprehend what is
dismissed as subjective emotional and irrational public sentimentality. Yet while public sentiment over protecting wilderness and endangered species defies objective science-based evaluation, it is real. Such
sentiment is part of our humanity, and to lose it or discount it is to
render our species inhuman, callous and unfeeling. It is from our sentimental sensibility that we appreciate and protect the life and beauty
of the world, enjoy the wonder of Nature, the innocence of infants. It is
the template of our conscience, against which we judge our own actions and that of others. Certainly sometimes sentiment can be mis-
placed and misguided, but such feelings as horror, outrage, and
sympathy are subjective elements of our morality, ethics, and perhaps
of a deeper intuitive wisdom. Public sentiment, therefore, should not
be discounted or manipulated by those who think they know better
and who claim to have the authority of science, God, or reason.

Biosophy and Bioethics: Seeds of Hope

There are too many people in the world for all of us to live simply
off the land. The land has been changed too much, as has its bounti-
fulness. We are all dependent upon technologies that can provide us

with fuel, food, clothing and shelter, health, and entertainment. These aspirations and expectations drive the industrial machine, be it capitalist, socialist, or fascist.

Without a biosophical basis, industrialism creates its own nemesis through entropy. This we are learning as members of the post-industrial age. It is a time of choice, of crisis, and opportunity. We can find great hope in the blessings that life has to offer us, but we must lower our expectations and demands upon all that sustains our lives. We must also practice one of the most charitable of all acts, which in earlier times spiritual leaders called chastity. Today that means making our families smaller, and adopting the children of the landless and the poor, whom the affluent and the privileged must help in developing sustainable self-sufficiency.

In sum, bioethics chastens us to live simply, modestly, and with dignity. Biosophy inspires us to repair, recycle, and reduce our needs and wants, to waste not and want not, to support organic farming and holistic medicine, to regulate family size, urban development, and the oil, timber, mining, and fishing industries. It also inspires us to integrate communications and transportation systems with equitable and efficient exchange of information and resources, while promoting the teaching of social justice, sustainable economics, and cultural and natural history in schools and colleges.

These innovations that many people long before me have advocated (Bill Mollison's permaculture movement is an example) are seeds of hope.[25] We must all, within the scope of our own lives and lifetimes, work as individuals and networks to help these seeds germinate. It is all a matter of choice and will. The future we choose we will help create today. Will tomorrow bring us any closer to the future that we hope for, for the good of our children, or for our sunset years? Or is that future incompatible with the survival of the meadow lark, the spotted owl, the wildebeest, and the two billion landless children who will be born this year? Until we realize the basic truth of biosophy, that all of life is interconnected, then the good that good people strive to do will continue to cause more harm than good, and our best intentions will amount to nothing. As Henry David Thoreau wrote in 1849, "If . . . the machine of government . . . is of such a nature that it requires you to be an agent of injustice to another, then, I say, break the law."[26] "Most of the luxuries and many of the so called comforts of life," Thoreau contended, "are not only not indispensable, but positive hinderances to the elevation of mankind."

Moralistic, economistic, scientistic, legalistic, and other simplistic interpretations of and solutions to the myriad problems societies face today are widespread and often destructive. The consequences of simplistic thinking, good intentions notwithstanding, are too often pernicious. This could be avoided by acknowledging that most problems are complex, that there may be more than one cause and one solution, and that no one has a monopoly on truth. It is regrettable that Pope John Paul II so vehemently asserts the infallibility of his view that sexual abstinence is the only acceptable method of family planning. He confuses celibacy with chastity and seems to demonstrate a lack of compassionate understanding of human nature. Over population and the poverty associated with environmental degradation are major causes of human conflict, suffering, and inhumanity today. Increasing crimes of violence, disease, and the erosion of moral restraint and ethical sensibility are symptoms, not causes.

Certainly the symptoms need to be dealt with, but the first medicine is prevention and this entails identifying and correcting the causes. The major source of the ills of the world have to do with human behavior, particularly our needs, wants, expectations, and attitudes toward others. We are a consciously transformative species with the innate capacity to change how we think and act. As history informs, we have not effectively begun to acknowledge the urgency of self-transformation into fully humane beings. This is now a survival imperative. Every human community must become humane, environmentally conscious, and self-sustaining. Communities can help others develop a humane sustainable society. Similarly, nation must help nation, since competition and conflict will quickly fill the vacuum where there is neither compassion nor cooperation.

From Harmfulness to Harmlessness:
The Ten-Percent Solution

It is said that in order to find the right answers, we must first ask the right questions. But these cannot be formulated when we are not in our right minds in the first place. When we lack the broad conceptual framework of bioethics, and a deep empathic concern and respect for all life, the right questions will not be forthcoming. Neither will we find the appropriate answers or solutions to the moral, social, and environmental problems that we face today. While it is more widely acknowledged today that we face a serious global environmental and economic crisis, the root cause of this crisis is not yet widely recognized, and in many circles it is even denied. This root cause is our state of mind, our attitude toward life, and our self-definition. Collectively, we regard ourselves as the rulers of Earth's Creation, seeing ourselves as separate from Nature rather than a part of Nature.

Our collective inhumanity toward animals and the life community of this planet is a symptom of a state of mind that is as diseased and dysfunctional as the world it creates.[27] We should be as concerned about the state of our own humanity, the human spirit, as we are about the holocaust of the animal kingdom and the desecration of the natural world. And we need to see that human salvation, animal liberation, and environmental restoration are codependent and one and the same. In recognizing that every being has a purpose, we shift from a humano-centric to an ecocentric worldview. The interconnectedness and interdependence of animal and human rights and environmental integrity demonstrates what Imam Al-Hafiz B. A. Masri and his nephew Nadeem Haque call the *equigenic* principle. When human society is in accord with this universal principle, it is humane, socially just, and sustainable.[28]

The laws of Nature are written all around us and were inscribed long before the first humans became part of the living history and "dream of the Earth."[29] The first human communities learned to live according to these laws. These first people, like the aboriginals of Australia and the people of the African bush, still live by these laws today and strive to survive against the onslaught of those who have no respect for the laws of Nature (or for the rights of indigenous peoples), and who fabricate their own laws to further self-interest and justify the rape of the Earth. Natural law teaches that every being serves a purpose. But when we pervert that purpose, or telos, to satisfy our own ends, as by eliminating some species and making others more abundant, we transgress this law. Our Indo-European ancestors transgressed this law 8–10,000 years ago when they began to commoditize a few varieties of domesticated plant and animal species. In the process, they began the annihilation of the natural world along with annihilation of the first people, aboriginal gatherer-hunters who never developed an economy based on the wholesale commoditization of other life forms.[30] The remedy for this biblically documented "fall" is for us to learn to farm in Nature's image, humanely and equitably with respect for wildlife, developing agricultural practices that are sustainable and conservation-oriented, thus enhancing biodiversity and cultural diversity.

Our history did not begin with the advent of agriculture or with the industrial revolution. All our roots go back to a far more ancient time during which our species evolved in relative harmony with the

Earth.[31] Our ancestors caused less harm than we descendants, because they were fewer in number, and lacked the technology and the desire to commoditize life. While we cannot turn back the clock and live as they did, we should not become so historically amnesic that we forget or deny the wisdom of our ancestors, the first people, or the folly and arrogance of the founding fathers of industrialism. Rather, we must evolve spiritually and socially, mindful of the fact that the *anima mundi*, or soul of the Earth, is part of us, and when we harm or destroy it, we do no less to ourselves. When we revere and renew it, the mystery and divine purpose of human existence will be revealed and celebrated. The alternative, evolutionary choice is to be an exterminator species, that brings natural history to an end, and exterminates our spirit, if not also our body.

I have heard the haunting "biomusic" of the south Indian jungle Kurumba tribals in the Nilgiris with whom I have played my flutes and didjeridoo, and seen the remains of elephants killed by poachers and electrocuted by rich land owners intent on protecting cash crops raised to launder "black" money. Both the Kurumbas and the Asian elephants are likely to become extinct soon because of the combined appetites of poverty and overpopulation, power and greed. Lacking the feral vision, and Nature wisdom arising from the immediate Earth-connected economy and way of life of these Indian and other indigenous peoples, those in civilized society, born into the worker-consumer industrialized nexus of capitalism and technological and military imperialism, take the "Evil Empire" as normative. Hence, they neither see it for what it is, nor do they see obedience to natural law as enlightened self-interest. Through the educational process, media manipulation of truth and control of information, and our desensitization and denial, the status quo of consumerism and industrialism is maintained. But for how long, we must ask, can this status quo be maintained, the harmful consequences of which must be seen as evil, because it is not sustainable? It will take the rest of the world down with it unless the bioethical seeds of our humanity, along with the seeds of biocultural diversity and of a sustainable agriculture and world economy, are protected from extinction.

Notes

Introduction

1. Van Rensselaer Potter (1971) *Bioethics, Bridge to the Future.* Englewood Cliffs: Prentice Hall.

2. Van Rensselaer Potter (1988) *Global Bioethics: Building on the Leopold Legacy.* East Lansing: Michigan State University Press.

3. George H. Kieffer (1979) *Bioethics: A Textbook of Issues.* Reading: Addison-Wesley.

4. Thomas Gladwin (1995) cited by Rick Clugston in "A paradigm shift in academic knowledge." *Earth Ethics* 7: No. 1, pp. 13–14.

Chapter 1

1. J. S. Stein and L. Urdang (eds) (1996) *The Random House Dictionary of the English Language: The Unabridged Edition.* New York: Random House.

2. See: M. W. Fox (1997) *Concepts in Ethology: Animal Behavior and Bioethics.* Malabar: Krieger Publishing Co.

3. Stein and Urdang.

4. C. Gilligan (1982) *In a Different Voice.* Cambridge: Harvard University Press.

5. Ch'u Ta-kao (1973) (Transl.) *Tao Te Ching.* New York: Samuel Weiser. p. 4.

6. Bill Neidjie (1991) *Speaking for the Earth: Nature's Law and the Aboriginal Way.* Washington DC: Center for Respect of Life and Environ-

ment; see also: M. W. Fox (1996) *The Boundless Circle*. Wheaton: Quest Books.

7. C. H. Turner (1971) *Radical Man*. New York: Anchor.

8. D. Brower (1965) (ed) *Not Man Apart: Lines from Robinson Jeffers*. New York: Ballantine, p. 34.

9. R. C. Austin (1985) *Environmental Ethics* 7:197–208.

10. E. Fromm (1956) *The Art of Loving*. New York: Harper & Row.

11. F. Dostoevsky (1981) *The Brothers Karamazov*. New York: Bantam Books.

12. Ch'u Ta-Kao. p. 30.

13. Gia-Fu Feng and J. English (1972) (Transl.) *Lao Tsu, Tao Te Ching*. New York: Random House. pp. 13 & 19.

14. K. Lorenz (1963) *On Aggression*. London: Methuen.

15. F. de Waal (1996) *Good Natured: The Origins of Right and Wrong in Humans and Other Animals*. Cambridge: Harvard University Press.

16. V. R. Potter (1995) "Global bioethics: Linking genes to ethical behavior." *Perspectives in Biol. & Med.* 39:118–31.

17. H. Skolimowski (1993) *A Sacred Place to Dwell: Living with Reverence Upon the Earth*. Rockport: Element Books.

18. R. Lockwood and F. R. Ascione (eds) (1998) *Cruelty to Animals and Interpersonal Violence*. West Lafayette: Purdue University Press.

19. Potter (1995) Global bioethics.

20. See: M. W. Fox (1996) *The Boundless Circle: Caring for Creatures and Creation*. Wheaton: Quest Books.

21. A. Maslow (1968) *Toward a Psychology of Being*. New York: Van Nostrand Reinhold.

22. Confucius, *The Great Learning*. Cited in Wolfe Lowenthal (1991) *There Are No Secrets*. Berkeley: North Atlantic Books. p. 73.

23. M. W. Fox (1978) "What future for man and earth? Toward a biospiritual ethic," in R. K. Morris and M. W. Fox (eds) *On the Fifth Day: Animal Rights and Human Ethics*. Washington: Acropolis Books Ltd, pp. 219–30; see also: G. Russo (1995) *Evangelium Vitaae. Commento all Enciclica Sulla Bioetica*. Torino, Italy. Editrice Elle Di Ci 10096 Leumann; and R. Otowicz (1998) *The Ethics of Life: The Bioethical and Theological Context of the Problem of Conceived Life*. Krakow: Wydawinctwo WAM. Ksieza Jezuici.

Chapter 2

1. T. A. Shannon (ed) (1993) *Bioethics: Basic Writings on the Key Ethical Questions that Surround the Major, Modern Biological Possibilities and Problems*. Fourth edition. Mahwah: Paulist Press;

H. T. Engelhardt, Jr. (1996) *The Foundation of Bioethics*. New York: Oxford University Press;

B. Gert, C. M. Culver, and K. D. Clouser (1997) *Bioethics: A Return to Fundamentals*. New York: Oxford University Press;

R. Devries and J. Subedi (eds) (1998) *Bioethics and Society: Constructing the Ethical Enterprise*. Upper Saddle River: Prentice Hall;

A. R. Jonsen (1998) *The Birth of Bioethics*. New York: Oxford University Press;

A. R. Jonsen, R. M. Veatch, and L. Walters (eds) (1998) *Source Book in Bioethics: A Documentary History*. Washington: Georgetown University Press;

A. Ridley (1996) *The Origins of Virtue: Human Instincts and the Evolution of Cooperation*. New York: Viking;

A. Ridley (1998) *Beginning Bioethics: A Text with Integrated Readings*. New York: St. Martin's Press.

2. For further discussion, see: B. E. Rollin (1981) *Animal Rights and Human Morality*. Buffalo: Prometheus Books; F. de Waal (1996) *Good Natured: The Origins of Right and Wrong in Humans and Other Animals*. Cambridge: Harvard University Press; D. DeGrazia (1996) *Taking Animals Seriously: Mental Life and Moral Status*. New York: Cambridge University Press; and, R. Nash (1989) *The Rights of Nature: A History of Environmental Ethics*. Madison: University of Wisconsin Press.

3. V. R. Potter (1970) "Bioethics, the science of survival." *Perspectives in Biology and Medicine* 14:127–53; see also: V. R. Potter (1970) "Biocybernetics and survival." *Zygon* 5: 229–46; V. R. Potter (1971) *Bioethics: Bridge to the Future*. Englewood Cliffs: Prentice Hall; and, W. T. Reich (1994) "The word bioethics: its birth and the legacies of those who shaped it." *Kennedy Institute of Ethics Journal* 4:319–35.

4. T. Berry (1988) *The Dream of the Earth*. San Francisco: Sierra Books.

5. E. O. Wilson and S. R. Kellert (eds) (1993) *The Biophilia Hypothesis*. Washington: Island Press; see also: D. Abram (1996) *The Spell of the Sensuous*. New York: Pantheon Books.

6. M. W. Fox (1990) *Inhumane Society: The American Way of Animal Exploitation*. New York: St. Martin's Press.

7. Potter (1988) *Global Bioethics*; see also: V. R. Potter and P. J. Whitehouse (1998) "Deep and global bioethics for a livable third millennium." *The Scientist*. January 5, p. 9.

8. J. Raloff (1998) "Drugged waters. Does it matter that pharmaceuticals are turning up in water supplies?" *Science News* 153:187–89.

9. P. Kropotkin (1955) *Mutual Aid: A Factor of Evolution*. Boston: Extending Horizons Books.

10. E. L. Ericson (1985) *The Free Mind Through the Ages*. New York: Frederick Ungar Publishing Co.

11. H. E. Daly and K. N. Townsend (eds) (1993) *Valuing the Earth: Economics, Ecology and Ethics*. Cambridge: The MIT Press; see also: E. Goldsmith (1992) *The Way: An Ecological View*. London: Rider; and David Korten (1995) *The Tyranny of the Global Economy*. West Hartford: Kumarian Press, Inc.

12. L. Eiseley (1969) *The Unexpected Universe*. New York: Harcourt, Brace and Jovanovich. p. 32.

13. H. E. Daly (1990–91) "Sustainable growth: a bad oxymoron." *Journal Envir. Health Sci.* C8(2).

14. V. R. Potter and L. Potter (1995) "Global bioethics: converting sustainable development to global survival." *Medicine and Global Survival* 2:185–90.

15. P. Kennedy (1993) *Preparing for the 21st Century*. New York: Random House; see also: H. Küng (1993) *Global Responsibility: In Search of a New World Ethic*. New York: Continuum Publishing Co.

16. Potter and Potter (1995) "Global bioethics."

17. See: Fox (1997) *Concepts in Ethology*.

18. A. MacIntyre (1984) *After Virtue*. Second edition. Paris: Notre Dame University Press; see also: T. L. Beauchamp and J. F. Childress (1994) *Principles of Biomedical Ethics*. 4th edition. New York: Oxford University Press; E. D. Pellegrino and D. C. Thomasma (1993) *The Virtues in Medical Practice*. New York: Oxford University Press; and W. T. Reich (ed) (1995) *Encyclopedia of Bioethics*. Second edition. New York: Macmillan.

19. See: J. Mason (1993) *An Unnatural Order: Uncovering the Roots of Our Domination of Nature and Each Other*. New York: Simon and Schuster.

20. See: M. W. Fox (1992) *Superpigs and Wondercorn: The Brave New World of Biotechnology and Where It All May Lead*. New York: Lyons and Burford.

21. See: R. W. Gerard (1989) "Underground allies of plants." *USDA's Agricultural Research*. November, pp. 10–13.

22. A. Leopold (1970) *A Sand County Almanac*. New York: Ballantine. p. 6.

23. E. O. Wilson (1994) *Biophilia*. Cambridge: Harvard University Press. p. 22.

24. A. Schweitzer (1947) *Albert Schweitzer: An Anthology*. New York: Harper & Bros. p. 79.

Chapter 3

1. A. Leopold (1970) *A Sand County Almanac*. pp. 238–239.

2. C. Stone (1974) *Should Trees Have Standing? Toward Legal Rights for Natural Objects*. Los Altos: William Kaufmann.

3. C. Stone (1987) *Earth and Other Ethics: The Case for Moral Pluralism*. New York: Harper & Row.

4. L. Margulis (1998) *The Symbiotic Planet: A New Look at Evolution*. New York: Basic Books.

5. H. Rolston III (1988) *Environmental Ethics: Duties and Values in the Natural World*. Philadelphia: Temple University Press. p. 187.

6. H. Rolston III (1989) *Philosophy Gone Wild: Environmental Ethics*. Buffalo: Prometheus Books. p. 91.

7. See: E. C. Hargrove (ed) (1992) *The Animal Rights / Environmental Ethics Debate: The Environmental Perspective*. Albany: State University of New York Press.

8. R. Attfield and A. Belsey (eds) (1994) *Philosophy and the Natural Environment*. Cambridge: Cambridge University Press; J. B. Calicott (1989) *In Defense of the Land Ethic*. Albany: State University of New York Press; R. F. Nash (1989) *The Rights of Nature: A History of Environmental Ethics*. Madison: The University of Wisconsin Press; and, P. Taylor (1986) *Respect for Nature: A Theory of Environmental Ethics*. Princeton: Princeton University Press.

9. T. Benton (1993) *Natural Relations: Ecology, Animal Rights and Social Justice*. London: Verso; B. Norton (1991) *Toward Unity Among Environmentalists*. New York: Oxford University Press; and, G. E. Varner (1998) *In Nature's Interests? Interests, Animal Rights and Environmental Ethics*. New York: Oxford University Press.

10. A. Naess (1973) "The shallow and deep, long-range ecology movement: A summary." *Inquiry* 16:95–100.

11. W. DeVall and G. Sessions (1985) *Deep Ecology: Living as if Nature Mattered*. Salt Lake City: Peregrine Smith Books. p. 65.

12. E. Abbey (1985) *The Monkey Wrench Gang*. Salt Lake City: Dream Garden Press; see also: D. Foreman (1991) *Confessions of an Eco-Warrior*. New York: Harmony Books.

13. B. R. Taylor (ed) (1995) *Ecological Resistance Movements: The Global Emergence of Radical and Popular Environmentalism*. Albany: State University of New York Press.

14. J. Davis (ed) (1991) *The Earth First! Reader*. Salt Lake City: Peregrine Smith Books; and, R. Gottlieb (1993) *Forcing the Spring: The Transformation of the American Environmental Movement*. Washington: Island Press.

15. G. Francione (1995) *Animals, Property and the Law*. Philadelphia: Temple University Press.

16. D. Henshaw (1989) *Animal Warfare: The Story of the Animal Liberation Front.* London: Fontana; R. Garner (1993) *Animals, Politics and Morality*. Manchester: Manchester University Press; H. Guither (1998) *Animal Rights: History and Scope of a Radical Social Movement*. Carbondale: Southern Illinois University Press.

17. L. Vikka (1997) *The Intrinsic Value of Nature*. Rodopi: Atlanta.

18. J. B. Callicott (1994) *The World's Great Ecological Insights: A Critical Survey of Traditional Environmental Ethics from the Mediterranean Basin to the Australian Outback*. Berkeley: University of California Press.

19. J. McDaniel (1989) *Of Gods and Pelicans: A Theology of Reverence for Life*. Louisville: Westminster/John Knox Press.

20. J. Lovelock (1979) *Gaia*. Oxford: Oxford University Press.

21. R. Attfield (1983) *The Ethics of Environmental Concern*. Oxford: Blackwell.

22. C. Birch and J. B. Cobb, Jr. (1990) *The Liberation of Life*. Denton: Environmental Ethics Books; J. E. Carroll, P. Brockelman and M. Westfall (eds) (1997) *The Greening of Faith: God, the Environment and the Good Life*. Hanover: University of New Hampshire; R. S. Gottlieb (ed) (1996) *This Sacred Earth: Religion, Nature and Environment*. New York: Routledge; and, D. G. Hallman (ed) (1994) *Ecotheology: Voices from South and North*. Maryknoll: Orbis Books.

23. J. Passmore (1974) *Man's Responsibility for Nature*. New York: C. Scribner's Sons.

24. *The Quest Magazine* (1999) "Ecological dharma." May/June. p. 101; see also: H. Skolimowski (1992) *Living Philosophy: Eco-philosophy as a Tree of Life*. London: Penguin; H. Skolimowski (1994) *Participatory Mind: A New Theory of Knowledge and of the Universe*. London: Penguin; and M. W. Fox (1984) *One Earth, One Mind*. Malabar: Krieger Publishing.

25. J. Moltmann (1990) "Human rights: the rights of humanity and the rights of nature." In: *The Ethics of World Religions and Human Rights*. H. Küng and J. Moltmann (eds). London: SCM Press. pp. 120–35.

26. S. McDonaugh (1986) *To Care for the Earth*. London: Geoffrey Chapman; and S. McHague (1993) *The Body of God*. London: SCM Press.

27. D. Griffin (ed) (1988) *The Re-enchantment of Science*. Albany: State University of New York Press.

28. A. Huxley (ed) (1958) *The Perennial Philosophy*. London: Fontana. p. 9.

29. M. Bookchin (1990) *The Philosophy of Social Ecology: Essays in Dialectical Naturalism*. New York: Black Rose Books.

30. R. Radford Ruether (1993) *God and Gaia: An Ecofeminist Theology of Earth Healing*. London: SMC Press.

31. C. Merchant (1980) *The Death of Nature: Women, Ecology, and the Scientific Revolution*. San Francisco: Harper & Row.

32. C. J. Adams (1994) *Neither Man Nor Beast: Feminism and the Defense of Animals*. New York: Continuum; see also: C. J. Adams and J. Donovan (eds) (1992) *The Animal Rights/Environmental Ethics Debate: The Environmental Perspective*. Albany: State University of New York Press; and, C. J. Adams and J. Donovan (eds) (1995) *Animals and Women: Feminist Theoretical Explorations*. Durham: Duke University Press.

33. J. Donovan and C. J. Adams (eds) (1996) *Beyond Animal Rights: A Feminist Caring Ethic for the Treatment of Animals*. New York: Continuum; see also: C. J. Adams (ed) (1993) *Ecofeminism and the Sacred*. New York: Continuum; G. Gaard (ed) (1993) *Ecofeminism: Women, Animals and Nature*. Philadelphia: Temple University Press; and, M. Spiegel (1988) *The Dreaded Comparison: Human and Animal Slavery*. London: Heretic Books.

34. R. Ryder (1983) *Victims of Science: The Use of Animals in Research*. London: Davis-Poynter.

35. P. Singer (1990 *Animal Liberation*. Second edition. New York: Avon Books.

36. T. Regan (1983) *The Case for Animal Rights*. Berkeley: University of California Press.

37. B. Rollin (1981) *Animal Rights and Human Morality*. Buffalo: Prometheus Books; S. R. L. Clark (1984) *The Moral Status of Animals*. New

York: Oxford University Press; A. Linzey (1987) *Christianity and the Rights of Animals*. New York: Crossroads; M. Midgley (1984) *Animals and Why They Matter*. Athens: University of Georgia Press.

38. H. Salt (1892) *Animal Rights*. Reprinted edition 1980. Clarks Summit: Society for Animal Rights; A. Schweitzer (1965) *The Teaching of Reverence for Life*. New York: Holt, Rinehart and Winston.

39. D. DeGrazia (1996) *Taking Animals Seriously: Mental Life and Moral Status*. New York: Cambridge University Press; M. Bekoff and D. Jamieson (1991) "Reflective ethology, applied philosophy and the moral status of animals." *Perspectives in Ethology* 9:1–47; see also an excellent reference resource on animal issues: M. Bekoff (ed) *Encyclopedia of Animal Rights and Animal Welfare*. Westport: Greenwood Press; S. C. Bostock (1993) *Zoos and Animal Rights*. London: Routledge; L. Finsen and S. Finsen (1994) *The Animal Rights Movement in America: From Compassion to Respect*. New York: Twayne; R. Fouts and S. Mills (1997) *Next of Kin*. New York: William and Morrow & Co.

40. M. Midgley (1994) "The end of anthropocentrism." In: R. Attfield and A. Belsey (eds) *Philosophy and the Natural Environment*. Cambridge: Cambridge University Press. p. 111; see also: M. Midgley (1994) *The Ethical Primate: Humans, Freedom and Morality*. London: Routledge.

41. B. Norton (1991) *Toward Unity Among Environmentalists*. New York: Oxford University Press. p. 246.

42. S. Clark (1994) "Global religion." In: R. Attfield and A. Belsey (eds) *Philosophy and the Natural Environment*. Cambridge: Cambridge University Press. p. 126; see also: S. Clark (1993) *How to Think About the Earth*. London: Mowbray.

43. J. B. Callicott (1980) "Animal liberation: A triangular affair." *Environmental Ethics* 2:311–38.

44. J. B. Callicott (1989) "Animal liberation and environmental ethics: Back together again." In: *Defense of the Land Ethic*. Albany: State University of New York Press. pp. 49–59.

45. J. B. Callicott (1994) *The World's Great Ecological Insights: A Critical Survey of Traditional Environmental Ethics from the Mediterranean Basin to the Australian Outback*. Berkeley: University of California Press. p. 52; see also: J. B. Callicott, et al., in: M. E. Tucker and J. M. Grim (eds) (1993) *Worldviews and Ecology*. Lewisburg: Bucknell University Press; and, S. R. Kellert (1996) *The Value of Life: Biological Diversity and Human Society*. Washington: Island Press.

46. B. Norton (1991) *Toward Unity Among Environmentalists*.

47. M. W. Fox (1983) "Philosophy, ecology, animal welfare, and the 'rights' question." In: H. B. Miller and W. H. Williams (eds) *Ethics and Animals*. Clifton: Humana Press. pp. 307–15.

48. M. W. Fox (1978) "What future for man and earth? Toward a biospiritual ethic." In: R. K. Morris and M. W. Fox (eds) *On the Fifth Day: Animal Rights and Human Ethics*. Washington: Acropolis Books, Ltd. pp. 219–30; see also: M. W. Fox (1984) *One Earth, One Mind*. Reprint edition. Malabar: R. E. Krieger Publishers.

49. T. Berry (1988) *Dream of the Earth*. San Francisco: Sierra Books.

50. See for further discussion: M. W. Fox (1998) *Concepts in Ethology: Animal Behavior and Bioethics*. Second edition. Malabar: R. E. Krieger Publishing Co.

51. D. Hume (1969) *A Treatise on Human Nature*. Oxford: The Clarendon Press.

52. C. Darwin (1904) *The Descent of Man and Selection in Relation to Sex*. New York: J. A. Hill & Co.

53. S. Clark (1997) *Animals and Their Moral Standing*. New York: Routledge.

Chapter 4

1. N. Myers (1979) *The Sinking Ark*. New York: Pergamon.

2. G. Hardin (1977) *The Limits of Altruism: An Ecologist's View of Survival*. Bloomington: Indiana University Press.

3. See "Interview with Alberta Thompson." *Earth First!* June/July 1999. p. 7.

Chapter 5

1. A. Richard Mordi (1991) *Attitudes Toward Wildlife in Botswana*. New York: Garland Publishing Co.

2. For a well-documented discussion of this subject, see: John A. Hoyt (1994) *Animals in Peril: How "Sustainable Use" Is Wiping Out the World's Wildlife*. Garden City: Avery.

3. For further discussion on the inherent risks of world free trade, see Chapters 9 and 12.

4. A. Schweitzer (1961) *Out of My Life and Thought: An Autobiography*. New York: Holt, Rinehart and Winston. pp. 158–9.

5. R. Frazer Nash (1989) *The Rights of Nature: A History of Environmental Ethics*. Madison: University of Wisconsin Press.

6. S. R. Kellert and E. O. Wilson (1993) *The Biophilia Hypothesis*. Washington: Island Press. p. 12.

7. C. Ruffensperger and J. Tickner (1999) (eds) *Protecting Public Health and the Environment: Implementing the Precautionary Principle*. Washington: Island Press.

8. John Terbough (1999) *Requiem for Nature*. Washington: Island Press.

Chapter 6

1. According to the U.N. Food and Agriculture Organization, *FAO Production Yearbook 1990* (Rome 1991), some 800 kg. of grain is used to feed livestock in the U.S. to meet the annual per capita average consumption of 42 kg. of beef, 20 kg. of pork, 44 kg. of poultry, 283 kg. of dairy products and 16 kg of eggs.

2. According to a 1986 report by the federal Office of Technology Assessment, animal diseases cost U.S. agriculture $17 billion annually (*Feedstuffs*. March 14, 1994).

3. Various domestic crop, food, and beverage industry by-products (like citrus pulp and brewer's grains) do play an important role in providing feed for integrated livestock and poultry production.

4. Fox (1997) *Eating with Conscience*. Troutdale: New Sage Press.

5. Figures from *Illinois Agricultural News* (1994) July 15th. p. C2.

6. R. J. Kuezmarski, et al. (1994) "Increasing prevalence of overweight among U.S. adults." *J. Amer. Med. Assoc.* 272:205–11 (according to an NBC news report, July 19, 1994).

7. Thomas Merton (1966) Statement published in *Unlived Life: A Manifesto Against Factory Farming*. London: Campaign Against Factory Farming.

8. Paul E. Waggoner (1994) *How Much Land Can Ten Billion People Spare for Nature?* Ames: Council for Agriculture, Science and Technology. p. 2.

9. T. Sanders and S. Reddy (1994) "Vegetarian diets and children." *Am. J. Clin. Nutr.* 59:(Suppl)1176s–81s.

Chapter 7

1. R. Hubbard and E. Wald (1999) *Exploding the Gene Myth: How Genetic Information Is Produced, Manipulated by Scientists, Physicians, Employers, Insurance Companies, Educators and Law Enforcers*. Boston: Beacon Press.

2. Utzi Mahnaimi and Marie Colvin (1998) "Israel planning ethnic bomb as Saddam caves in." *The London Times*. Sunday edition. November 15.

3. For an excellent discussion on this subject, see: Richard Strohman (1994) "Epigenesis: The missing beat in biotechnology?" *Bio / Technology 12*: 156–64.

4. Over 4,000 experimental releases of genetically engineered organisms, mainly plants, had been approved by the U.S. government as of August 1999. The accidental release of genetically engineered (GE) fish from commercial fish ponds, the deliberate release of GE bacteria or insects to control crop pests, and the spread of pollen from GE crops that can result in genetic contamination of conventional and organic crops and wild relatives are cause for concern.

5. Mae-Wan Ho (1998) *Genetic Engineering: Dream or Nightmare?* Bath: Gateway Books.

6. For further discussion on this important issue, see: Vandana Shiva (1993) *Monocultures of the Mind: Biodiversity, Biotechnology and the Third World*. Penang: Third World Network; and Bernard Rollin (1995) *The Frankenstein Syndrome: Ethical and Social Issues in the Genetic Engineering of Animals*. New York: Cambridge University Press.

7. T. J. Hoban and P. A. Kendall (July 1992) *Consumer Attitudes About the Use of Biotechnology in Agriculture and Food Production*. Washington: USDA Extension Service.

8. The sale of pesticides, fertilizers and new hybrid seeds under the aegis of aid and development programs to industrialize agriculture in the third world helped destroy more sustainable, traditional agricultural practices, caused extensive ecological and economic damage and resulted in much poverty and malnutrition as peasants were driven off their land to make way for large agricultural enterprises. See: Edward Goldsmith (1992) *The Way: An Ecological World View*. London: Rider, and Addendum.

9. Quotation from *Bio / Technology*. (1993) 11: p. 13.

10. Fox (1992) *Superpigs and Wondercorn*.

11. The U.S. has not yet become a party to this convention on biodiversity because some Senators fear it would jeopardize American industry, even

though President Clinton added his signature to it. See: Andrew Pollack (1999) "U.S. and allies block treaty on genetically altered goods." *The New York Times*, February 25.

12. T. Roszak (1992) *The Voice of the Earth*. New York: Simon & Schuster.

13. T. Berry (1988) *The Dream of the Earth*. San Francisco: Sierra Books.

14. Fox (1992) *Superpigs and Wondercorn*. p. 169.

15. S. Donnelley, C. R. McCarthy, and R. Singleton, Jr. (1994) *The Brave New World of Biotechnology*. Hastings Center Report, Special Supplement 24:1. p. 14.

16. S. Donnelly et al. (1994) p. 15.

17. S. Donnelly et al. (1994) p. 22.

18. On the topic of agriculture, see: M. W. Fox (1996) *Agricide: The Hidden Farm and Food Crisis That Affects Us All*. Second edition. Malabar: Krieger Publishing Co. On the topic of medicine, see: B. Inglis (1981) *Diseases of Civilization*. Seven Oaks: Hodder & Stoughton Ltd.

19. K. Dawkins (1997) *Gene Wars: The Politics of Biotechnology*. New York: Seven Stories Press; see also: V. Shiva (1997) *Biopiracy: The Plunder of Nature and Knowledge*. Boston: South End Press.

20. Fox (1992) *Superpigs and Wondercorn*; see also: D. Suzuki and P. Knudtson (1990) *Genethics: The Clash Between the New Genetics and Human Values*. Cambridge: Harvard University Press; and J. Doyle (1985) *Altered Harvest*. New York: Viking.

21. Fox (1997) *Eating with Conscience*.

22. N. Myers (1998) *Perverse Subsidies: Tax Dollars Undercutting Our Economies and Environments Alike*. Winnipeg: International Institute for Sustainable Development.

23. J. Lazarou, B. H. Pomeranz, and P. N. Corey (1998) "Incidence of adverse drug reactions in hospitalized patients." *JAMA*, April 15, Vol. 279, No. 15, pp. 1200–5.

24. N. D. Barnard, A. Nicholson, and J. Lil Howard (1995) "The medical costs attributable to meat consumption." *Preventive Medicine* 24:646–55.

25. See: N. J. Temple and D. P. Burkitt (1994) *Western Diseases: Their Dietary Prevention*. Totowa: Humana Press; and J. Chesworth (ed) (1996) *The Ecology of Health: Identifying Issues and Alternatives*. Thousand Oaks: Sage Publications Ltd.

26. M. W. Fox (1999) *Beyond Evolution: The Genetically Altered Future of Plants, Animals, the Earth, and Humans*. New York: The Lyons Press.

27. Declaration of the international movement for ecological agriculture. (1990) From: "Global crisis towards ecological agriculture." *The Ecologist 21*:107–12.

Chapter 8

1. See: Benjamin Farrington (1949) *Francis Bacon: Philosopher of Industrial Science*. New York: Henry Schuman. p. 5.

2. Farrington p. 91.

3. Farrington p. 58.

4. Farrington pp. 148–9.

5. It is worth reflecting on the notion that the coded information in the genes that form a double helix of DNA (deoxyribonucleic acid) "is the word of God made flesh."

6. B. Klug (1983) "Lab animals, Francis Bacon and the culture of science." In: *Listening, Journal of Religion and Culture* 18:54–72.

7. Farrington pp. 182–3.

8. J. F. Haught (1986) "The emergent environment and the problems of cosmic purpose." *Environmental Ethics* 8:139–50.

9. R. Levins, R. Lewontin (1985) *The Dialectical Biologist*. Boston: Harvard University Press. p. 208.

10. Science has nothing to do, either, with faith—defined by J. MacMurray as a principle of valuation by means of which a man decides what is worthwhile and what is not. Science thus cannot provide a faith for the modern world. It can only provide the means for achieving what we want to achieve. See: John MacMurray (1968) *Freedom in the Modern World*. Second edition. London: Faber and Faber. p. 36. (First published in 1932).

11. M. Berman (1981) *The Reenchantment of the World*. Ithaca: Cornell University Press. p. 16.

12. Berman p. 23.

13. A. Drengson (1980) "Shifting paradigms: From the technocratic to the person-planetary." *Environmental Ethics* 3:222–40; see also: David Ehrenfeld (1978) *The Arrogance of Humanism*. New York: Oxford University Press.

14. T. Tulka (1977) *Time, Space and Knowledge: A New Vision of Reality*. California: Dharma Press. p. 73.

15. Excerpted from *On Social Concerns* (Sollicitudeo Rei Socialis) *Chap. 4, Encyclical letter of John Paul II*. December 30, 1987.

16. F. E. Winkler (1960) *Man, The Bridge Between Two Worlds*. New York: Harper and Brothers. p. 158 and p. 250.

17. R. Dubos (1972) *A God Within*. New York: Charles Scribner's. p. 39.

18. Albert Einstein (1954) *Ideas and Opinions*. Carl Seelig (ed) New York: Crown Publishers. p. 11.

19. A. Schweitzer (1965) *The Teaching of Reverence for Life*. New York: Holt, Rinehart and Winston. p. 93.

20. MacMurray p. 75.

21. *Ibid*. p. 216.

22. J. Grim (1981) "Reflections on Shamanism." *Teilhard Studies 6*: Fall. White Plains: American Teilhard Association for the Future of Man, Inc.

23. It is unfortunate from a spiritual perspective, but perhaps understandable, that the Pope should speak of "kingship" rather than "kinship," and perpetuate the dualistic worldview of spirit and matter being separate. See also: Report in *Science* (1980), 207:1165–69.

24. Martin Luther King (1983) *Quotations from His Works*. New York: Newmarket Press. pp. 63 and 67.

25. C. P. Snow (1964) In: *The Two Cultures: A Second Look*. Cambridge: Cambridge University Press.

26. F. Crick (1966) *Molecules and Men*. Seattle: University of Washington Press. pp. 93–94.

27. W. Berry (1977) *The Unsettling of America: Culture and Agriculture*. San Francisco: Sierra Club Books. pp. 137–138.

28. L. White (1967) *Science 155*:1203–07. I do not entirely agree here with White, since the influence of Rome on Christianity and of Greek rationalism, particularly St. Thomas Aquinas, who embraced Aristotle's worldview, is considerable.

29. The cognitive and perceptual mind-set which, for example, sees *Homo sapiens* as something separate from and superior to the rest of Creation.

30. A return to traditional values does not imply a regression to "primitive" ways.

31. W. N. Perry (1986) *A Treasury of Traditional Wisdom*. San Francisco: Harper & Row. p. 49.

32. G. Bateson (1975) *Steps to an Ecology of Mind*. New York: Ballantine. p. 269.

33. W. Leiss (1974) *The Domination of Nature*. Boston: Beacon Press p. 3.

34. G. Himmelfarb (1996) In: *The De-Moralization of Society. From Victorian Virtues to Modern Values*. New York: Knopf. p. 12.

35. Mae-Wan Ho (1998) *Genetic Engineering: Dream or Nightmare*. Bath: Gateway Books. Also for further documentation, see also: M. W. Fox (1999) *Beyond Evolution: The Genetically Altered Future of Plants, Animals, the Earth, and Humans*. New York: The Lyons Press.

36. G. B. Adams and D. L. Balfour (1998) *Unmasking Administrative Evil*. London: Sage Publications Ltd.

Chapter 9

1. E. F. Schumaker (1973) *Small Is Beautiful: Economics as if People Mattered*. New York: Harper & Row.

2. P. Kropotkin (1955) *Mutual Aid: A Factor of Evolution*. Boston: Extending Horizons Books.

3. P. Teilhard de Chardin (1964) *The Future of Man*. New York: Harper & Row.

4. Fox (1996) *The Boundless Circle*. Wheaton: Quest Books.

5. B. Neidjie, *Speaking for the Earth* (1991) Washington, DC: Center for Respect of Life and Environment.

6. Cartesianism denotes the philosophy of René Descartes whose notions like the separation of mind and body, the supremacy of reason over emotion, and the idea that animals are unfeeling automata gained wide acceptance in the 18th century and endure to this day.

7. B. Swimme and T. Berry (1992) *The Universe Story*. San Francisco: Harper. p. 243 and 260.

8. P. Teilhard de Chardin op. cit.

Chapter 10

1. Fox (1992) *Superpigs and Wondercorn*.

2. H. E. Daly and K. N. Townsend (eds) (1993) *Valuing the Earth: Economics, Ecology, Ethics*. Cambridge: The MIT Press; see also: Paul Hawken, L. Hunter Lovins and Amory Lovins (1999) *Natural Capitalism:*

Creating the Next Industrial Revolution. New York: Little, Brown and Co.; and Paul Hawken (1994) *The Ecology of Commerce: A Declaration of Sustainability*. New York: Harper Collins.

3. H. E. Daly (1989) In: *Earth Ethics*. Fall. Washington, DC: Center for Respect of Life and Environment. p. 6.

4. United Nations (1987) *Our Common Future*. New York: UN World Commission on Environment and Development.

5. E. Goldsmith (1992) *The Way: An Ecological Worldview*. London: Rider. p. 6.

6. D. Worster (1977) *Nature's Economy*. San Francisco: Sierra Club Books.

7. R. Radcliffe-Brown (1972) *Structures and Function in Primitive Society*. London: Cohen and West.

8. P. Kropotkin (1955) *Mutual Aid: A Factor of Evolution*. p. 3.

9. J. McDaniel (1990) *Earth, Sky, Gods and Mortals: Developing an Ecological Spirituality*. Mystic: Twenty-third Publications.

Chapter 11

1. Ivan Illich (1976) *Medical Nemesis*. New York: Random House; see also: M. W. Fox (1986) *Agricide: The Hidden Crisis that Affects Us All*. New York: Schocken Books.

2. *The Washington Post* (1994) Dec. 17th. p. A22.

3. See: C. M. Turnbull (1968) *Forest People*. New York: Touchstone; and C. M. Turnbull (1972) *Mountain People*. New York: Touchstone.

4. Originally published in the *Hastings Center Report*, March/April 1989, pp. 40–2.

Chapter 12

1. See: L. M. Singhvi (1990) *The Jain Declaration on Nature*. New Delhi: Bhagwan Mahavir Memorial Samiti. Also: Padmanabh S. Jaini (1979) *The Jaina Path of Purification*. Berkeley: University of California Press.

2. M. K. Gandhi (1959) *My Socialism*. Ahmedabad: Navjivan Publishing House. pp. 34–5.

3. Knut A. Jacobsen (1994) "The institutionalization of the ethics of 'non-injury' toward all 'beings' in ancient India." *Environmental Ethics* 16:287–301.

4. While organically certified produce may cost more, because organic farmers don't receive many government subsidies, local marketing cooperatives and direct purchasing from farmers (so-called Community Supported Agriculture) can significantly reduce prices. For details, see: M. W. Fox (1997) *Eating with Conscience: The Bioethics of Food*. Troutdale: NewSage Press.

5. In these monotheistic traditions there is a dualism or Manichaean split between matter and spirit; between divinity and Earth's Creation; and between the temporal, existential self and the eternal self. This helped lay the foundation for industrial age secular materialism, atheistic scientism, and imperialism.

6. The doctrine of reincarnation affirms the immortality of the souls of all living beings. There is no single physical existence and then an eternity in some heaven or hell. Rather, the soul is manifested in various physical forms as it evolves along a predetermined path toward self-realization and liberation into the Absolute.

Chapter 13

1. J. Corbett (1991) *Goatwalking: A Guide to Wildland Living*. New York: Viking. p. 23 and p. 192.

2. Corbett p. 193.

3. D. Korten (1994) "Sustainable livelihoods: redefining the global social crisis." *Earth Ethics*. Fall. p. 11.

4. Corbett pp. 13–14.

5. J. B. Judis. "America Divided" in *Book World. The Washington Post*, p. 11, January 15, 1995. A review of Christopher Lusch's *The Revolt of the Elites*.

6. D. Korten (1995) *The Tyranny of the Global Economy*. West Hartford: Kumarian Press Inc. p. 102.

7. *Regulatory Policy Guidelines* (1983) No. 5, p. 19. Presidential task force chaired by Vice President George Bush. Washington: Govt. Printing Office.

8. Count L. N. Tolstoi (1889) Trans. by Huntington Smith, *My Religion*. London: Walter Scott.

Chapter 14

1. D. R. J. Macer (1994) *Bioethics for the People by the People*. Christchurch: Eubios Institute.

2. A. Gore (1992) *Earth in the Balance*. New York: Houghton & Mifflin.

3. A. Maslow (1968) *Toward a Psychology of Being*. New York: Van Nostrand Reinhold.

4. For further discussion, see: F. Bormann and S. R. Kellert (eds) (1991) *Ecology, Economics, Ethics: The Broken Circle*. New Haven: Yale University Press; D. Ehrenfeld (1993) *Beginning Again: People and Nature in the New Millennium*. New York: Oxford University Press; J. R. Engel and J. G. Engel (eds) (1990) *Ethics of Environment and Development: Global Challenge, International Response*. London: Belhaven Press; and, T. Rozak (1978) *Person / Planet: The Creative Disintegration of Industrial Society*. Garden City: Doubleday.

5. See: Bruce Rich (1994) *Mortgaging the Earth: The World Bank, Environmental Impoverishment and the Crisis of Development*. Boston: Beacon Press; see also: D. Goulet (1995) *Development Ethics: A Guide to Theory and Practice*. New York: Apex Press; IUCN (1995) *International Covenant on Environment and Development*. Bonn: World Conservation Union (IUCN), Commission of Environmental Law; IUCN, UNEP, and WWF (1991) *Caring for the Earth: A Strategy for Sustainable Living*. Gland: World Conservation Union (IUCN), United Nations Environment Programme (UNEP), and World Wide Fund for Nature (WWF); A. Naess (1989) *Ecology, Community and Lifestyle*. (Translation from the Norwegian and revision by D. Rothenberg) Cambridge: Cambridge University Press; and, United Nations (1993) *Agenda 21: Programme of Action for Sustainable Development: Rio Declaration on Environment and Development: Statement of Forest Principles*. New York: United Nations, Document No. DPI/1344 (April).

6. For further discussion, see: P. R. Ehrlich and A. H. Ehrlich (1991) *Healing the Planet: Strategies for Resolving the Environmental Crisis*. New York: Addison-Wesley Publishing Co.; and, G. Hardin (1993) *Living Within Limits: Ecology, Economics and Population Taboos*. New York: Oxford University Press.

7. D. R. J. Macer et al. (1994) *Bioethics for the People by the People*. Christchurch: Eubios Ethics Institute. p. 7.

8. Macer op cit p. 8.

9. William Drozdiak (1994) Poor nations resist tougher trade rules. *The Washington Post*. April 14. p. 8.

10. M. K. Gandhi (1959) *My Socialism*. Ahmedabad: Navajivan Publishing House. p. 36.

11. (1964) *Collected Works of Mahatma Gandhi*, Vol. XIII. Delhi: Government of India. p. 219.

12. M. K. Gandhi (1966) *The Village Reconstruction*. Bombay: Bharatiya Vidya Bhavan. p. 30.

13. E. F. Schumacher (1965) "Buddhist economics." In: *Asia: A Handbook*. Ed: Guy Wint. London: Anthony Blond. p. 699; see also: E. F. Schumacher (1975) *Small Is Beautiful*. New York: Perennial/Harper & Row.

14. R. Nader et al. (1998) *The Case Against "Free Trade": GATT, NAFTA, and the Globalization of Corporate Power*. San Francisco: Earth Island Press.

15. W. Berry (1998) in: R. Nader et al., op cit. p. 231.

16. Macer op cit., p. 29.

17. G. Orwell (1984) *Nineteen Eighty Four*. Reprint edition. New York: Penguin Books. (Original edition was printed in 1949.) See also: D. Ehrenfeld (1978) *Arrogance of Humanism*. New York: Oxford University Press.

18. H. E. Daly, J. B. Cobb, and C. W. Cobb (1994) *For the Common Good: Redirecting the Economy Toward Community, the Environment, and a Sustainable Future*. Second edition. Boston: Beacon Press.

19. *The Washington Post* (1994) May 18th p. A4.

20. *The Economist* (1994) May 24th p. 6.

21. For further discussion, see: W. Berry (1978) *The Unsettling of America: Culture and Agriculture*. New York: Avon/Sierra Club Books.

22. D. Macer (1994) "Bioethics may transform public policy in Japan." *Politics and the Life Sciences*. February, pp. 89–90.

23. C. Chapple (1993) *Nonviolence to Animals, Earth and Self in Asian Traditions*. Albany: State University of New York Press. p. 9.

24. I. Illich (1977) *Medical Nemesis: The Expropriation of Health*. New York: Bantam.

25. B. Mollison and D. Holmgren (1981) *Permaculture One: A Perennial Agriculture for Human Settlements*. Winter: Tagari Books/International Tree Crops Institute.

26. Henry David Thoreau (1996) *Walden and Civil Disobedience*. Owen Thomas (ed) New York: W. W. Norton. p. 231 and p. 9.

27. For an excellent critique on science's attitude toward animal pain and consciousness, see: B. Rollin (1989) *The Unheeded Cry: Animal Consciousness, Animal Pain and Science*. New York: Oxford University Press; see also: M. W. Fox (1990) *Inhumane Society: The American Way of Exploiting Animals*. New York: St. Martin's Press.

28. *Nature in Islam: The Concern for Ecological Balance* (unpublished manuscript); see also: Al-Hafiz B. A. Masri (1987) *Islamic Concern for Animals*. Petersfield: The Athene Trust.

29. J. Campbell (1983) *The Way of the Animal Powers, Volume One: Historical Atlas of World Mythology*. San Francisco: Harper & Row, Alfred Van DerMarck editions; J. Campbell (1988) *Historical Atlas of World Mythology, Volume Two: The Way of the Seeded Earth, Part One: The Sacrifice*. New York: Harper & Row; and, V. Deloria, Jr. (1983) *God Is Red*. New York: Dell.

30. M. Gimbutas (1974) *The Gods and Goddesses of Old Europe, 7000 to 3500 B.C.: Myths, Legends and Cult Images*. London: Tims and Hudson; and Berkeley: University of California.

31. L. Vanderpost and J. Taylor (1985) *Testament to the Bushmen*. New York: Penguin Books.

Index